SILVER BULLETS…

AND OTHER ENERGY EFFICIENCY MYTHS AND MAGIC!

BY: ANDREW VAILLENCOURT

EDITED BY: RICHARD R. VAILLENCOURT,
ASEE, BSME, PE, LEED AP

Copyright © 2016 by Andrew Vaillencourt

The Slide Rule Group, LLC
25 Mortimer Rd
Moosup CT, 06354

Ordering Information:
Special discounts are available on quantity purchases by corporations, associations, and others. For details, contact the publisher at the address above.

Orders by U.S. trade bookstores and wholesalers. Please contact:

Andrew@TheSlideRuleGroup.com

<u>Dedication</u>

For Harold, who taught me to pick fights.
For Mom, who taught me that it was OK to pick fights.
For John, who taught me to enjoy picking fights.
For Dad, who taught me how to win the fights.

Table of Contents

Silver Bullets... and Other Energy Efficiency Myths and Magic!

"One man's magic is another man's engineering." –Robert Heinlein

Energy Conservation has, in a relatively short period of time, become one of the most important factors in driving the evolution of industry in the modern world. Where it was once the incremental concern of large industries with massive processes now even the mildest shopkeeper or office manager must be concerned with the when's, where's, and how's of their respective energy loads. Kilowatts and BTU's are no longer solely relegated to the conversations of bespectacled nerds and utility company employees. In this modern world, it seems nobody can escape the looming specter of rising energy costs.

As is typically the case, this newly-minted problem in a previously under-served paradigm has given birth to an entire industry of inventors, purveyors, and consultants. This motley crew of professionals stand devoted to helping the poor overwrought consumer to minimize the inflating cost of energy in business. And since all these individuals are supremely qualified, hyper-scrupulous, and incapable of error in either judgment or analysis we can confidently and happily buy whatever they are selling, right?

I guess not, huh?

There is a concept we as humans have come to have great emotional attachment to: We like "silver bullets." This is a cognitive bias toward easy low-investment solutions in the face of complex problems.

Taking the time to understand and evaluate every problem is counterproductive to getting through the day. One cannot simply stop and research entirely new fields of thought and endeavor every time a random issue pops up that requires a solution. Very literally, our brains have evolved to seek out and prioritize simplicity in problem-solving over deep thought and careful evaluation.

When your house is on fire, you do not care about the vagaries and subtleties of combustion. If you already know that water extinguishes fire, then that is the sum total of what you NEED to understand to effect a solution to the problem.

Would a better understanding of combustion make you a better firefighter? Would it open up new potential solutions? Absolutely. But advanced understanding is not NECESSARY to fix the problem. Then again, fire is not particularly complicated.[1]

Let's talk about werewolves for a moment. In the sense of the accepted folklore, we are talking about a highly complex problem. We have a ferocious and malicious creature that can instantly heal wounds, readily hide amongst the populace, and spread its curse willy-nilly. When presented with the problem of a feral lycanthrope, the protagonist in our fanciful hypothetical here does not have any real chance of success...unless our hero possesses 200 or so grains of silver. Suddenly, the big, bad wolf is dispatched with little muss or fuss at all. There you have it! A huge, complicated, and terrifying problem is solved with nice, neat, and convenient little solution. Ninety minutes later we are walking out of the theater satisfied with a story well-told and indigestion from bad popcorn.

Silver bullets in this context refer to easy, intuitive, and low-engagement solutions to complex problems. Our brain likes these, because understanding complicated systems is

---

1 Except for when it is complicated. Which can be surprisingly often!

hard, and simple solutions are easy. We don't have to understand how a problem started or why it is pervasive if we already have the easy solution for it in our hands. In a pattern that will rapidly become obvious to you, this is the point in the narrative where I ruin everything by pointing out the problem:

The real world rarely affords us the luxury of a silver bullet.

Complicated problems usually have complicated solutions or no real solutions at all.

Which brings us to Energy Efficiency. How buildings and processes use energy is a massively complex and thorny morass of industry-specific needs and the nearly infinite potential combinations of equipment configurations. Was that a lot of syllables? Let me clarify: Buildings are unique. Industries are unique. People are unique. Not every problem has a common solution. No one device or technique will work all the time. Figuring out what solution your particular problem needs is a big, hairy, complicated, and time-consuming process.

The problem with big, complicated problems is that it is really hard to mass-produce solutions. Nevertheless, in the interests of building successful products and companies the Energy Efficiency industry will often try to do just that. They will happily sell you mass-produced cartons of silver bullets. More often than not, you are buying a fantasy. This is fine if you are attending a good werewolf movie, but not when you are dealing with real world problems.

There are very few silver bullets out there, folks. Plenty of good old–fashioned lead bullets are available, though. They are not as sexy, and they are often expensive and difficult to master. But in the hands of good shooter...they work just fine.

Which is not to say that every company trying to save you energy is trying to scam the consumer. What it means is that in many cases the salesman may not truly understand what a facility needs. What he has is a hammer, what he sees when he looks at the customer is a nail.[2] Our goal is to not be the nail. The nail gets hammered. Not the fun kind of hammered, either.

Furthermore, one does not need to be an "engineer" to understand when something may not work. The word "engineer" will pop up a lot in this text. Don't get too hung up on it. It refers to the various species of technologically savvy individuals that saturate the efficiency product landscape. There are many flavors of engineer, ranging from a "sales engineer" or a "solutions engineer" and sometimes even "energy engineer." Be advised, that as long as the person is not actually practicing within the classic engineering disciplines, there is no certification for the term "Engineer." Electrical, Mechanical, Chemical, Civil, Structural, and Biomedical Engineering all require certification and endorsement by various governing bodies to employ the term "engineer." There are many and myriad types of engineer out there that do not enjoy such scrutiny, and "Energy Engineer" is among them.

This does not mean that engineers practicing outside of the classic definitions are not competent, it just means that the discerning consumer would be best served to evaluate their offerings on their own merit. This appeals to me because quite frankly, you should be doing this no matter who is presenting it. Stacks of degrees and certifications be damned, there is nothing magical about evaluating projects. All you need is a fairly decent understanding of

---

[2] *Some may take this time to point out that many engineers and technical consultants will display a certain "anti-salesman" bias. We assure you that this is not the case. We hate them without any bias whatsoever. It is a pure hate, driven by the cold, unflinching certainty of the scientific method and backed by the historical record.*

how your building uses energy and some very basic thermodynamics. Call it "practical engineering." A little bit of "practical engineering" goes a long way and does not require differential equations or advanced calculus.

Practical engineering not only looks at the pure science. Pure science is important for many things, none of which we are going to talk about in this text. Practical engineering looks at how those scientific principles are applied, where they are applied, and why they are (and are not) applicable to the specific situation in question. The ability to comprehend the origins of the universe and the subtleties of its primal forces will be largely unnecessary.

This is important! We want the Magic, but Magic can turn into Myth through myth-application (sorry, couldn't resist).[3] When someone invests time and money hoping for magic, receiving the mythical can embitter our poor customer to any future conservation measures. As an industry, Energy Efficiency has always been well-served by healthy skepticism and robust review, but when bitter people become jaded, nobody wins.

So how can this humble text help? In this book, we will attempt to provide structure and techniques for evaluating vendor claims, as well as discuss the most pervasive yet misapplied measures one is likely to see as a building operator. We are going to pull back the curtain and help you to see the most common "myths" of the efficiency industry, and maybe help you do a little "magic" of your own!

---

[3] *This awful pun is entirely attributable to Richard Vaillencourt, PE. Andrew Vaillencourt takes no responsibility for it or any consequences associated with it.*

What Could Go Wrong?

Experience tells us that the efficiency industry is typically plagued with three similar but distinct species of myths:

1. *The Myth of the Intuitive:* The first type myth deals with energy conservation measures that are flawed but look like they must work. The subtle lure of the "intuitive" in complex measures and techniques will often lead both vendors and purchasers astray. If a measure "makes sense" how can it be wrong?

2. *The Myth of the Quantifiable:* The second type of myth deals with conservation measures that are demonstrably effective. They are true engineering "magic" in that they can absolutely work. The myth is not whether the specific measure will produce energy savings; rather the myth lies in the assertion that anyone can calculate accurately what those savings will be.

3. *The Myth of Magnitude:* This myth combines the first two in that sometimes a measure will actually produce savings, but that the magnitude of those savings is nowhere near what it appears it should be on the surface.

Promises are tough things for the (good) vendor. Even when confidence in an effect is high, the circumstances around said effect can still be difficult to predict. As an engineer, one can state confidently that the sun will come up in the morning. If forced, we may even postulate that it will do so in the cardinal direction generally accepted as "East." But an engineer cannot assure you that you will be able to see it! Sometimes it's just too cloudy.

The effects of Energy Conservation Measures (ECMs) are like sunrises. In many cases we know what should happen but can't lock down all the detail. Other times we can build fairly robust predictive models that are visually impressive and get all the math right. Even under the best circumstances there is still no assurance that the system or measure will perform the way it has been modelled. There are simply too many variables. This is where myths and magic become indistinguishable. Many efficiency myths are born without guile or malice. They appear because too much confidence is given to the model before all the system parameters are understood.

This text will attempt to address individual energy saving strategies as they are presented in the marketplace and discuss their actual operation and performance in the real world. Not only is it important to determine if the savings will be real, it is just as important to be able to calculate the predicted savings with a degree of confidence appropriate to the methodology. Ultimately, if success is measured by whether we can "see the sun rise in the morning," then we must accept responsibility of getting above the clouds.

A Note on Stuff in Boxes

Throughout this book, you will come upon sections of text in boxes...much like this one. These sections contain the tech-heavy, hyper-specific, bone-dry and often really complicated[4] background data or industry knowledge. The insights in this book are for everyone and it is intended be just as useful to the layman as it is the experienced building operator or engineer.

It is important to understand where the reasoning that drives all of this comes from, and that is why these sections exist. What is not important is that you understand all the nuance and heavy calculus that goes with it. Honestly, I don't understand every little bit of it either. I hate calculus

---

[4] *"Boring"*

and calculus hates me.[5] Most vendors do not understand the really hard science nearly as well as they want you to think they do, anyway.

So, I encourage you to read through these segments and try to grasp as much as you can. But if you finish some of these sections and think to yourself, "What the hell is he talking about?" Don't worry. It's not critical that all of that stuff make perfect sense to you. What is important is that responsible people show their work and cite their references.

And if you do understand all of this stuff perfectly well, please don't point out my errors to anyone!

---

[5] *There was an "incident."*

Concepts and Definitions

Glossary of Terms

Let's get these out of the way now, shall we? There will be many words tossed around in this book and by vendors. You do not need a masters-level comprehension of them, but if you put some effort into familiarizing yourself with them now, you will be much harder to fool later.

ACH- (Air changes per Hour)

A measurement of the rate at which outside air infiltrates a building. One ACH is defined as outside air infiltration equal to the interior volume of said building per hour.

AHU- (Air handling unit)

Any mechanical device that conditions air and moves it to occupied space. Typically, an AHU will consist of at least one fan, and one heating or cooling coil.

BIN hours or 'Temperature bins'

BIN hours refer to the practice of parsing out a location's annual hours into temperature-specific 'bins' for the purposes of calculating energy loads. Typically done over a 30-year average, a location will have a finite number of hours at any given temperature. These hours get grouped into 5-degree bins.

For example: Hartford, CT experiences (on average) 18 annual hours between 90 and 95 degrees. Ergo there are 18 bin-hours in the 92.5-degree bin. Hartford's 'BIN data' looks like this (emphasis on the 92.5-degree bin):

OA Temp	# Hours
97.5	0
92.5	18
87.5	109
82.5	274
77.5	403
72.5	519
67.5	876
62.5	1069
57.5	804
52.5	698
47.5	810
42.5	771
37.5	823
32.5	588
27.5	262
22.5	267
17.5	203
12.5	107
7.5	88
2.5	58
-2.5	13
-7.5	0
-12.5	0

BTU- (British Thermal Unit)

The quantity of thermal energy required to raise one pound of water one degree Fahrenheit at sea level.

BtuH- (British Thermal Unit per hour)

Rate of thermal energy delivery required to raise one pound of water one degree Fahrenheit at sea level per hour

CCF-(100 Cubic Feet)

100 Cubic feet of natural gas, typically containing 103,200 BTU heat energy.

CCU- (Compressor/condenser unit)

One half of a direct expansion (DX) split system. It contains the compressor and condensing portions of the unit. This is typically the part of the unit mounted outside the space to be conditioned.

CFM- (Cubic Feet per Minute)

Rate of air movement.

Cooling Ton-(Ton)

This is defined as the rate of heat transfer that results in the melting of 1 short ton(2,000 lbs.) of pure ice at 32 °F in 24 hours. It's also a nice even 12,000 Btu/hr. Most refrigeration and cooling systems will be rated in tons.

CV- (Constant Volume)

Type of air handling system that does not experience variations in system air volume.

DDC- (Direct Digital Control)

Controls system that employs digital signals to exert control over equipment, as opposed to pneumatic or manual control.

DX- (Direct expansion)

Cooling system employing compressed refrigerant to transfer heat from conditioned space to unconditioned space. Requires a condenser and an evaporator.

ECM- (Energy Conservation Measure)

A product, strategy, technique, or system installed, employed or implemented to effect a reduction in energy consumption.

ECM (motors)- (Electronically commutated motor)

Electronically commutated motors (ECMs, EC motors) are synchronous motors that are powered by a DC electric source via an integrated inverter/switching power supply, which produces an AC electric signal to drive the motor. We care because they are really efficient.

EER- (Energy Efficiency Ratio)

The ratio of *output* cooling (in Btu/hr.) to *input* electrical power (in watts) at a given operating point.

EMS- (Energy Management System)

Centralized controls architecture responsible for managing building equipment. Installed and operated for the purpose of controlling energy consumption. Said system can be digital/electronic or pneumatic. Also: BMS BCS, BAS.

ESCO- (Energy Services Company)

Purveyor of products and solutions pursuant to increasing the efficiency of its clients and customers.

Heat Pump

Heat engine that transfers heat from one area to another via alternately compressing and evaporating refrigerant. Essentially a DX split system where the refrigerant can flow either direction allowing for heating as well as cooling.

HVAC- (Heating, Ventilating, and Air Conditioning)

Acronym for describing equipment and/or industries relevant to the Heating, Ventilating, and Air Conditioning of buildings.

IBR- (International Boiler Rating)

Industry standard rating system for boilers. Codifies methodology for manufacturers of heating equipment.

MBtu-(1,000 BTU)

Can be either instantaneous or per hour.

MMBtu-(1,000,000 BTU)

Can be either instantaneous or per hour.

PTAC- (Packaged Terminal Air Conditioner)

Catch-all term for air handler mounted through an exterior wall and containing all necessary controls within the unit itself. PTACs are ubiquitously employed in hotel rooms.

RTU- (Rooftop Unit)

An AHU mounted on the roof of a building. Differentiated due to engineering challenges unique to rooftop mounting.

R-value

A numerical rating of the insulating properties of a substance. Equal to $1 \text{ h ft}^2 \cdot {}^\circ\text{F/Btu}$ or the inverse of U-factor.

Setback

ECM wherein space setpoints are altered during unoccupied periods to reduce energy consumption. Can be either scheduled or dynamic.

Therm

100,000 Btu. Colloquially (if incorrectly) used as unit of natural gas consumption.

Split System

DX cooling system with an indoor evaporator coil and a remotely located CCU.

U-factor

Heat transfer coefficient equal to the BTU per hour per square foot per degree (F) differential of a given material. (Really just the inverse of R-value).

VAV- (Variable Air Volume)

Air handling system wherein controls interact to vary the volume of air within the system to match loads. Lower air volumes mean less energy is required, making these systems inherently more efficient than constant volume.

VFD- (Variable Frequency Drive)

Motor control that alters the incoming frequency of alternating current cycles and subsequently yields power at varying frequencies. Since motor shaft speed is a linear function of current frequency, this allows the VFD to change the speed of a motor without mechanical assistance. Reducing the speed of a motor significantly reduces the power consumption.

VIV- (Variable Inlet Vane)

Fan intake vanes that alter the quantity of air available to a fan while at the same time acting as pre-rotation turning vanes thus reducing the pressure drop across the vanes. These were commonly used to create variable air volume systems before the advent of mass-produced VFD technology.

Useful Concepts

Throughout this text, we will reference many basic concepts and use some weird industry terms. These will be repeated often, and if internalized, can be applied to virtually any measure that is presented. If you are an experienced facility operator or energy professional, much of this may already be familiar to you.

Thermodynamic Continuity

Thermodynamics is not just a hairy physics concept that is simultaneously boring and complicated. It is a fundamental law of the universe, and so it must be addressed. Like gravity you can ignore it all you want, but you are still going to crash into the ground if you fall off that ladder. It's non-negotiable.

Here is a crash course in thermodynamics:

Thermodynamics is the part of physics that covers the relationship between heat and temperature and energy and work. The most important things to keep in mind are:

1. You cannot get more energy out of something than what you put in. If you want to get 10,000 BTU's out of a heater, you have to make sure that it has 10,000 (or more) BTU's in it. If the energy density of natural gas is 103,200 Btu/CCF, then the best possible output from any natural gas heater is 103,200 BTU per CCF. Realistically, it's less than that.

2. Energy always flows from high to low. Hot things get colder, fluids and gasses under high pressure will expand to lower pressures. We may employ a lot of devices to push energy around, but we do so by manipulating the conditions to get the energy to

flow where we want. If we want to make hot water, then we have to put the water somewhere warmer than it currently is (a boiler). In the boiler, the energy will flow form the (hot) flame to the (cooler) water.

3. Energy cannot be created or destroyed. IT MUST GO SOMEWHERE. 100% percent of the energy that comes down your wire or through your gas pipe goes somewhere. Heat lost from your boiler heats up the boiler room, the portion of electricity your lights don't turn in to light becomes heat, and machines give off heat and noise as they work.

Pure thermodynamics can get very complicated very quickly. Depending on your love of multivariable calculus you may or may not want to look at a given project as the sum of its various moving (thermodynamically) parts. The good news is that it is rarely necessary to do so.

In order to properly evaluate the effect of any specific conservation measure, you must discover and follow an unbroken thermodynamic connection (and the corresponding control connection) from the energy source, to the end use. Without exception, the end use will use or return all the energy that it's given. That's what makes it an end-use.

Take a room in a house. If the temperature of that room is to remain constant, then the rate of energy delivered to it must exactly equal the rate of energy that is lost. This is called "thermodynamic equilibrium" and that is the stated goal of every control system. If you deliver too much or too little energy, the space temperature will change until such time as equilibrium between the "energy in" and the "energy out" is achieved.

For the purposes of evaluating a product or measure, one can typically glean the necessary level of skepticism or optimism by evaluating arithmetically the chain of devices and effects between the points of origin (source) to the end use (sink). If you take the time to know and understand the devices that manipulate energy in your space, then this will not require a lot of time or heavy math.

Let's go ahead and heat a space, and follow energy from source to sink as we go:

1. We shall start with a boiler (source). The boiler converts fuel into heat by burning it. This flame heats water inside the boiler. The energy locked within the fuel has now been transferred to the water.

2. The hot water pump takes the hot water from the boiler and moves it to a radiator coil in the space to be heated. Some of the energy leaves the water through conduction and radiation and heats the air in the space.

3. After the coil, the (now-cooler) water returns the boiler to be heated up again and sent back through the coil.

4. This continuous flow of heating energy in the form of hot water through the coil causes he space to become warmer and warmer until the setpoint is reached.

5. The thermostat detects this and sends a signal to a valve in the radiator and tells it to close. Water stops flowing through the radiator, and heat energy stops flowing into the now-satisfied space (sink).

If the thermostat failed to close the valve, the space would continue to heat up until the amount of heat entering the room was equal to the heat leaving it (as mentioned, the much-vaunted "thermodynamic equilibrium."). It is the thermostat's job to control the energy flow into the space so that equilibrium occurs at the temperature you want it to. Too much energy and the room gets hot. Too little and it stays cold.

As far as any building is concerned supplying the energy that is needed to meet a load is the responsibility of system installed to do so. Typically, this system is designed and installed with capacity to deliver a much higher rate of energy than is expected to be required at the highest calculated load. It is the responsibility of the control system and equipment to limit the rate of energy transfer to only what is needed at any given time. Just because you *can* deliver five million BTUs an hour doesn't mean you *need* to. *The controls exist to match the system output to the load.* It's why you have a thermostat at all.

Understanding the continuity between the controls (in this case, the thermostat) and the systems that supply the energy is crucial to understanding how your building utilizes energy.

ΔT

Thermodynamics dictates that energy will always proceed from higher to lower (hotter to cooler); and never the other way around. Hot, delicious, life-affirming coffee on your desk will get cooler and more horrible as it sits. Since it is hotter than the air around it, the heat (energy) leaves the coffee to find its way into the air. Conversely, an ice-cold beer will become warm and disgusting as the heat from the warmer air migrates into its foamy goodness.

ΔT ("DELTA T") refers to the difference in temperature (energy) between any two things. Most of the time, we are either talking about water or air.

Let's get back to our beverages. If your space is 70°F, and your beer is 35°F, then the ΔT between beer and air is 35°F (70-35=35). In your building, if it is 70 inside, and 35 outside, then that ΔT is also 35°F. In the case of a heat engine, it is between the heat source and the heat sink (whatever they may be). I suppose instead of ΔT we could just say "the difference between," but that would be lame. Engineers think "ΔT" "sounds cooler." All this really does is illustrate how poorly engineers understand the concept of "coolness" as a social construct.

Efficiency

The definition of "efficiency" is the ratio of work input to the work output of a system. Practically, it boils down to "output stuff divided by input stuff." It is deceptively simple: if I have to pump 100 therms of natural gas into my boiler to get 85 therms of heat out of it, then this boiler is 85% efficient. Many vendors will throw this term around very casually, but the only acceptable formula is the one stated here. If used to describe anything other than that relationship, then it is time to get nervous about this salesperson.

Power

Power is the rate at which work is done. Apropos to your electric bill, it is the rate at which the utility has to jam electrons down the wire to meet your load. This is an instantaneous thing and has no time component. On your electric bill, this is the "kW" part, often described as "demand." It is NOT the "kWh" part of you bill. Don't let the vendor confuse you on this.

Head

"Head" means pressure; typically measured in feet. "Feet" refers to the amount of force generated by a column of water x feet above the ground. If you want to sound impressive in a meeting, you can drop lines like "that system puts out 65 feet of head at the outlet." Don't worry, no one will know what you are talking about, anyway.

Simple Payback

Simple Payback means the amount of time in years that it will take the savings from an ECM to pay for the installation of that ECM. If your ECM will save 100 dollars a year, and costs 1,000 to install, then the Simple Payback will be $1,000/$100 or 10 years. If you are an accountant, think of simple payback as the inverse of ROI (Return on Investment). Also, my condolences on being an accountant.

Loading

"Load" as a general term means the objective quantity of work that must be accomplished to achieve a goal.

When presented as a percentage, "Load" means how much of a system's available capacity is required to achieve the desired result. For instance, a boiler that is 85% loaded needs to output 85% of its total capacity to hold temperature in the space.

"Load" expressed as a non percentage quantity means how much work or energy is required to achieve thermodynamic equilibrium. (hold a setpoint.)

Most importantly; "load" is determined by the conditions the equipment is operating in and under. It is a function of the environment and is not something that can be altered without altering the environment.

What do I mean by this? An example:

If a space set to 70°F experiences 10,000 Btu per hour of heating *load* when it is 30°F outside, then the equipment serving this space must deliver that many Btu to meet that load (hold that setpoint).

You could change that 10,000 Btu load by adding more insulation to the walls, or turning down the setpoint, or otherwise changing the environment and conditions. You cannot change the load by changing the boiler.

Why do we care? We care because some vendors get confused and think that attaching a device to a boiler can change the load on it.

Load is not something that the equipment itself can change. A bigger boiler may be more lightly *loaded* as a percentage of capacity, but the *load* on the bigger boiler is the same 10,000 btu/hr. as it was on the smaller one.

The critical consideration here is that you can only save the energy that is *wasted* in the process of meeting the load.

If you need 10,000 btu every hour to hold setpoint, and your boiler is 85% efficient, then you will have to put 11,765 Btu/hr. through it to get the required 10,000 Btu out of it. If you had a more efficient boiler, then naturally you know you can save *some* of the extra 1,765 Btu that never made it to the space.

Now imagine that you have a vendor saying he can save you 20% (2,000 Btu/hr.) with his boiler widget. This is somewhat paradoxical, as you are only wasting 1,765 Btu in the first place. The customer is left to wonder how he can possibly save you more energy than is actually being wasted!

By the way, there are two solutions to that paradox: Either it won't save that much, or it will save that much and thus not meet the load

The Best Advice you will Ever Get

Before we go any further, there is one very important task that you should undertake before venturing into the inhospitable wasteland of energy efficiency products and vendors:

Know your building.

Even if you are not a plumber, or an electrician, or an engineer; you can still go and learn about your facility. What kind of heating does it employ? What type of cooling? What are the distribution systems? What are the limitations and liabilities of your occupants?

The less you understand the way your building is using energy, the easier it will be for a vendor to confuse you. Either inadvertently or intentionally, that vendor is coming at you with a hammer, remember? But you are not a nail. You are a 3/16th hex screw with a reverse thread.[6]

You need to communicate this to your vendors in a way that they will understand, which is not easy. It must be done to ensure that your vendor gets you the right products. Truthfully, most vendors are perfectly happy to be directed by you on this. They want happy customers, and they want return customers. If you have the knowledge and information they need to get you the right product, then they don't have to guess as to what you need.   Make no mistake, in the absence of good information, they *will* try to guess. You can't even hold that

---

[6] *This is the engineering equivalent to being a "special snowflake." Ironically, engineers do not accept that snowflakes are special. Since they are all unique, then there is nothing special about their uniqueness. Ergo, none are special. You are not special.*

You're welcome.

against them. They are not in the business of turning away customers. But if you can guide them they are almost universally delighted to help you get the right equipment for your goals. Bring a solid understanding of your needs to the interaction and everybody is going to be happier in the long run.

Which brings us to the next section...

Energy Audits: What They Are and What They Aren't!

Energy Audits are very en vogue right now. They are a prerequisite for any number of utility and government incentive programs, as well as most commercial building energy usage rating systems.

Unfortunately, the energy audit has suffered from the same issues as the ECMs themselves have since the advent of the efficiency industry boom. Namely, the need to mass-produce a saleable product to a rapidly expanding consumer base.

This creates a crisis of identity within the auditing community, as the nature of the audit and the product itself means many things to many people.

For clarity, the American Society of Heating, Refrigeration, and Air-Conditioning Engineers (ASHRAE)[7] classifies three distinct types of energy audit:

Type of Audit

Highlights

Level 1

- Rapid assessment of building energy systems
- Building energy benchmark
- High-level definition of energy system optimization opportunities
- Outline applicable incentive programs

Level 2

---

[7] *Runner up to "Federal Employee Antidiscrimination and Retaliation Act of 2002 (FEAR)" for "Least Helpful Acronym Ever" award.*

- Detailed building survey of systems and operations
- Breakdown of energy source and end use
- Identification of ECMs for each energy system
- Range of savings and costs for the ECMs
- Spotlight on Operational Discrepancies
- Outlining priorities for limited resources, next steps, and identification of ECMs requiring more thorough data collection and analysis (ASHRAE Level-3)

Level 3

- Longer term of data collection and analysis
- Accurate modeling of ECMs and power/energy response
- Bid-level construction cost estimating
- Investment-grade, decision-making support

All of which is great, but what does that mean for you? And what really happens in the field? Let me break it down for you:

- **Level One**
  - Auditor will walk through your building, ask a lot of questions, and then send you a report outlining in very broad terms your energy use and all the things the auditor feels might be worth exploring.
  - These are typically offered free by vendors to try to get you to let them in the door, so they can try to shoe-horn their product into your building.
- **Level Two**
  - The auditor will walk more slowly and take more notes. At this level, the auditor may

take spot readings of actual equipment (kW, amps, airflow, etc.)

- o Auditor will send a fairly detailed report with actual calculations with estimated costs and savings, and a fairly robust analysis of the ECMs identified.
- o Most building owners will get all the information they need to make an informed decision about ECMs at this level.
- o These are rarely free.
- Level Three
  - o This is the real deal. This is a comprehensive, investment-grade audit. There will be multiple visits, long-term data logging and many measurements. This will result in a very robust analysis. When a competent professional does a Level Three, you can typically have a lot of confidence in the results.
  - o These are not cheap. But a good one is worth every penny. These audits scare vendors.

The decision to have an audit done at your facility can be as simple as allowing a vendor to walk through and make their pitch, or as complicated as bringing in a consulting engineer to measure everything available.

When you are dealing with vendors, it is important to remember that knowledge is power. The more you know, the less that disreputable companies can get away with. When you are dealing with a trusted vendor, solid knowledge means that no one has to guess about what you need and what the potential is. You want to make sure that you have the best audit you can afford, and the best personal knowledge of your facility you can acquire.

This is your facility, and your responsibility.

The Most Common Tricks

Many of the erroneous claims made by vendors and purveyors of ECMs and products follow certain patterns. Spotting these patterns can go a long way towards avoiding a nasty surprise on your utility bill after an expensive installation. Like most of the gremlins that infest the good science and solid engineering of the industry, a lot of these are innocuous and understandable oversights. Some of them are simply aggressive sales tactics. A few are straight-up deceptive.

It's up to you to make a judgment call as to exactly how and why you believe your vendor is employing them.

For any project there are two major areas to look at: Baseline and Projected. "Baseline" refers to your current conditions, and "projected" refers to the results after you implement the ECM. The most common "spin error" is to inflate the Baseline. This will start you at a higher level of existing consumption than actually exists and therefore inflate the savings when you "fix" the problem. Several typical sales tactics use this *modus operendi*...

Let's look at a few of the more common tricks:

Too Many Run Hours

Energy consumption is, quite simply, load X time. If your device pulls 15kW for one hour, you get billed for 15 kWh (15kw X 1H = 15 kWh).

To inflate savings, the vendor may simply assume a very large number of run hours to begin with, giving the measure a larger pool of costs to save against. This is most common in lighting and VFD calculations. Adding an extra 1,000 run hours to a big lighting job can add thousands of (non-existent) dollars to the savings.

Data-logging devices are fairly cheap and easy to use. Go ahead and measure your run hours to see for yourself. Knowing your run hours beforehand can help your salesperson use the right number. Otherwise, they will guess, and probably guess high.

Too Much Loading

If inflating the run hours works, then inflating the load does as well. Assuming an 8 million BTU boiler is 100% loaded means assuming 8 million BTUS per hour for all those run hours. But, if that boiler is 100% oversized, then it is really only burning 4 million BTUs. That is a huge discrepancy. Much like run hours, figuring out your loading can prevent nasty surprises later.

The worst offenders for this trick are VFD projects and condensing boilers.

Big Losses

When calculating savings for ECMs that minimize any sort of system loss, assuming high losses is the easiest way to make the product look more attractive. Most system losses are less than what they may appear, and not quantifying these introduces error into the calculation.

Honestly, most types of system losses are notoriously difficult to quantify, so tread carefully here. Even the best vendor will likely be doing a lot of guesswork unless you pay for a very comprehensive audit.

This is most commonly associated with hot water systems, building insulation, or steam distribution.

Tanking the Combustion Efficiency

Most boiler/furnace ECMs center on improving the efficiency of the total system. Which makes perfect sense. Logically, it follows that the best way to save lot of money with these is if the existing piece of equipment is terrible. Here's the rub:

Boilers and furnaces don't have loads of moving parts. Mankind has been burning fuel for heat for about 200,000 years, and we have most of the kinks worked out. I have measured combustion efficiency on a 61-year-old natural gas boiler that had enjoyed a life comprised entirely of neglect, abuse, and indignity, and it still managed to be 82.4% efficient. I typically see vendors assuming combustion efficiency on 20-year-old boilers assumed at 70% or less. Which is not to say that bad boilers aren't out there, just that I would not let the vendor do their own testing. Your local boiler guy will do the test independently for a couple of hundred bucks. It's worth it.

This is most commonly associated with boilers, obviously.

Projection Inflection

Reducing the "projected" conditions is less common, simply because they will always be there, and are thus subject to real measurements. This frightens the disreputable vendor. Inflated baselines are more common because it they are easier to get away with. (If baseline inflation isn't suspected before installation, it won't be there to be measured after. By the time you realize a mistake was made, there will be no way to prove it.)

Optimistic "projected" conditions lower the level of expected consumption than what is likely to occur. This shows a larger quantity of calculated savings at the time you make your economic (and career) decision to implement the project.

But low projected usage happens primarily when the baseline is already well established. The vendor needs those savings to sell the project, so they are content to roll the dice with a low projected usage. Basically, take all the items form "Inflating the Baseline" that start with the words "too many" or "too much" and replace them with "too few" and "too little".

How will they get away with it? They will blame the victim, of course. When the project fails to produce the savings, the vendor then brings out his reasons why *you* were the problem and not their projections. Which is an easier tactic for them to get away with.

The most common excuse for the failure is that *you* didn't run the equipment the way that you agreed to run it. You don't remember agreeing? You agreed as soon as you accepted their calculations.

You should go over the calculations very carefully and ask questions whenever something is not clear to you. You don't have to ask these questions in front of your boss which would let him know that you don't understand this stuff. You can always arrange a private meeting or contract with a hired gun[8] to ask the questions.

Adherence to the "Rule of Thumb"

Rules of Thumb are helpful for quickly determining the feasibility of an idea. They are usually based upon the vendor's typical experience with a typical customer. For any discrete project, they are terrible indicators of actual potential. Vendors like them because they allow a company to put together a very impressive-looking package with a minimal amount of work. The issue is that they are wrong as often as they are right, and depending on the skill of the applier, they are often completely misapplied.

When it comes to laying out *your* money, you don't really care how it *usually* works anywhere else. Since the money is coming from *your* budget, you need to know how it will

---

8 *Twirls pistols theatrically...*

work at *your* building. That is a tougher question to answer, but a critical question to answer correctly.

This one is completely ubiquitous.

Technobabble

Some of the best ECMs out there rely on complicated physical principles. Some vendors exploit this by trying to dazzle you with highly technical language and extremely complicated charts, tables, and formulas. It's OK not to understand these, but it can be very difficult to know if you are being bamboozled. The trick to defeating this is to ask for all the documentation of the measure's potential and the vendor's claims and sagely tell the vendor you will look it all over.

Then, after the vendor leaves, go and furiously[9] google the measure in question until you find out what the hell is going on. There are a lot of internet forums and websites out there that cover many of these measures. You might have to dig a little, but the information is out there, and the search is worthwhile.

Or you can call up a qualified energy professional and pay them to evaluate the claims of the vendor. Its money well spent if you aren't sure what you are looking at.

Technobabble is most often associated with Power Conditioning Devices, but really applies to any vendor who doesn't know what they are doing.

Testimonials

Testimonials are the lowest form of validation. Testimonials are not science, they are emotion. Facts do not

---

[9] *The fury helps.*

require endorsement, and cherry-picking testimonials is about as hard as fabricating them. Which isn't hard.

Now, just because a vendor has testimonials and/or references doesn't make them bad. But *relying* on testimonials is the reddest of red flags. Having happy customers is very comforting and working with the kind of vendors that leave satisfied customers behind them is important. If a vendor is proud of their products and customer satisfaction there is no reason not to include that in a sales presentation. But if that is all they've got, run like mad.

Argumentum ad Vericundium

This means "Arguing from authority." It is the fancified Latin version of "because I said so." The assumption being that the speaker or the person the speaker is representing is an authority on the subject, and thus automatically right. With vendors, this manifests as "Our guys are really good," or "our model is really good," whenever you ask for clarification or quantification of the methodology or results.

If you find yourself trying to pry technical answers out of a vendor and you are not as tech savvy as they are, this one will show up a lot.

"That's Proprietary"

If you aren't actually asking to see their personal trade secrets, this one shouldn't fly at all. The laws of physics and thermodynamics are not trademarked by anyone. This canard should only apply to questions specific to the manufacturing process of the device. This person wants you to install a device in your building and around your people. If they can't tell you what's in it, how it works, why it works, and what it does, get rid of them.

Savings Attribution

This one is especially sneaky. It's when the vendor's device appears to be producing savings, but these savings are attributable to something else entirely.

Let's imagine that some fuel-saving widget shows savings from one winter to the next. So it's working, right? Not if the second winter is warmer than the first! It is very common for deficiencies in the heating system to be magnified when the winter is abnormally cold. Many people go out and buy fuel-saving device when this happens, and when the next year is warmer the bills go down, and they feel validated and their purchases...even though the bill would have gone down if they hadn't done anything!

Another version of this is when a vendor starts with some device that is poorly maintained or broken. If your equipment is in bad shape and performing poorly, the salesman can capitalize on this by promising that their device will increase your device's efficiency. Then, during installation the vendor repairs or services the equipment as part of the upgrade. The now-functional device ceases to do its job poorly, and savings materialize. Of course, just fixing the thing would have generated the same savings, but you will never know that. Instead of a low-cost maintenance call, you now have an expensive widget.

The Myths

On to some of the most common and specific technological faux pas of the energy efficiency marketplace!

Let's start with the worst offender. The biggest myth in this industry, the myth that persists and permeates everything Energy Consultants and Engineers do, is actually very insidious. Here is the unvarnished truth:

> *The consultant/engineer/vendor's job is not to get you to use less energy.*

Believing that less energy is always better is an easy trap to fall into. It is natural to assume the job of an energy engineer/consultant/manager etc. is to get you to use less energy. Why else would you want one?

No matter how intuitive or seductive that thought is, it is false. **It is the job of the energy professional to ensure that you only use exactly as much energy as is required to accomplish your goals.**

No more, no less. You operate a business that has a purpose and goals. Your building or facility exists for the sole purpose of achieving that end. Energy is a resource you must purchase in pursuit of this, but the goal is what is important.

Imagine you are the CEO of Global Framistats, Inc. Your goal might be to produce 1,000,000 of your patented Pileated Semi-Articulated Framistats a day. This is critical because Framistats are your big money-maker. You are the primary provider of Framistats for the whole North American continent, after all. If I come in and tell you that I can cut your energy bills in half, but that your production will go down by 50% then I really haven't really saved you anything.

I am a fraud. Throw me out.

If I am doing my job correctly, then you will more than likely hear me offering to cut your bill by 10%, *while preserving your output*. Getting more output product for less input resources is my job. That is the very definition of "efficiency." Reducing input is easy. We can just start turning things off. But maintaining your desired output while reducing the energy input? That's where the magic happens.

So now we shall get into the myths themselves. We will try to start with the easiest and progress to the more complicated ones. This text is designed to be accessible to all levels of building operator, but sometimes the thermodynamics get a tad hairy. Worse than that some of the electrical stuff makes very little sense to anyone, including electrical engineers[10].

Do not despair, the central concepts remain the same no matter how wonky the math gets. But keep this in mind at all times: misconceptions thrive when complicated things are portrayed too simply. It is a trap that has snared minds both brilliant and mediocre since the dawn of time and the advent of the salesman. For the sake of clarity and brevity, we shall attempt to only dive into the math as deeply as is required to understand where the myth is born.

And make no mistake, virtually every efficiency myth is born in a place of legitimate understanding. Very few of the various ECM's in this book are completely without merit, and that is because they are almost all based upon one very sound thermodynamic principle or another that had the misfortune to be convoluted at some point.

---

[10] *It's true. They will pretend it's not the case but ask them to do a "Fourier's Transformation" and watch them squirm.*

This text will focus on the most common HVAC and power ECMs, since actual manufacturing process is far too individualized to be addressed in this volume. The concepts, however, apply to all energy consumption. It does no matter if you are cooling molds, injecting high-density polyurethane, or thermally oxidizing volatile organic compounds. Nothing escapes the principles and fundamentals of energy transfer. Think of this book as "stage one" in developing your "Energy Conservation Consciousness," or something like that.

If you are just the property manager of facilities manager for an office building do not make the mistake of believing you can skip over these concepts, either. Even if you aren't running giant machines on three shifts you need to believe that your building is a box that you pour energy into, and the work done by the people inside is the product that comes out. Your box has leaks, and they are costing you productivity and money. Your job is to plug the leaks and only put as much energy in as you must.

Beware the myths enumerated herein, because they will harm you if you let them. Some of these myths are more egregious than others. Some of them are subtle and situational. All of them have a grain of truth at the center. That's what makes them compelling and pervasive. This is where we must become discerning consumers and preface our purchases with hard questions about how our buildings use energy. Only then we can alter those uses in a manner compliant with good science and good conscience.

Or I guess you could just buy whatever they are selling...

Myth #1: Tricks of the Light.

Lighting is the most popular ECM offered on the market today. Everybody will sell you new lights, and frankly, there are plenty of companies out there calling themselves ESCOs that will only sell you lights. Furthermore, many utilities will offer very attractive incentive packages to encourage you to change your lights.

This is a good thing because lighting upgrades do an awesome job of saving your facility energy. The average office building attributes between 20% and 40% of its total electric load to lighting This makes lighting upgrades the easiest, biggest, and most pervasive elephant in any energy-efficiency room. Modern improvements in lighting technology have reduced the energy requirement for lighting by 75% (or more) over the last twenty years, and this has caused a massive global push for facility upgrades.

Even better, lighting upgrades are not typically the most expensive improvement you can make to your building. With low-to-medium costs, and nice, high savings, lighting upgrade projects are the darling of the efficiency industry. As they should be. There is a lot of very good technology out there at a good price making great savings.

Enumerating those savings is simple and easy, as well. If Lamp[11] A pulls 100 watts, and lamp B pulls 50 watts, then you do not need to employ a lot of higher order math to determine that lamp B uses half as much juice as lamp A. Over 1,000 hours, the savings calculation will look like this:

---

[11] For the record, "lamp" refers to the part of the light fixture most people would call a "bulb." Light fixtures have lots of parts, but the "lamp" is where the light comes out of. It is not always bulbous, so it's not always a "bulb." Semantics is fun.

Existing (lamp A) = 100W X 1000h = 100,000 Wh = 100 kWh

Proposed (lamp B) = 50W X 1000h = 50,000 Wh = 50 kWh

Savings = 100kWh – 50kWh = 50kWh

If you can handle 4th grade math, you can do this.

So why is lighting in a book about myths? Because as the biggest seller (and therefore biggest money-maker) in the industry, a lot of shady practices have sprouted to help keep the sales numbers high. Such as...

Run Hours and the Myth of Magnitude

We touched on this one is an earlier section. Inflating the run hours is the most common trick in the industry for getting the savings to look better than they really are.

Take our previous example. We have one lamp that will save 50kWh over 1,000 run hours. Let's assume you have a big facility, and therefore you have 1,000 of these lamps.

So, your savings now look like this:

50kWh saved per fixture X 1,000 fixtures = 50,000kWh saved.

If you are paying $.12/kWh your cash savings is:

$.12 X 50,000kWh = $6,000

But uh, oh! The installation is going to be $25,000, which gives the project a 4.16-year payback. Naturally, the hate-filled nether-demons masquerading as accountants in your finance department insist that they will only allocate funds for jobs with a 2.5-year payback.[12]

Not a problem! Your trusty vendor will just go ahead and assume 2,000 run hours and double those savings! Now the payback is 2.1 years, and everybody wins. Of course, when

---

[12] *The quantity of corporate finance departments who have this exact ridiculous rule is staggering.*

your boss doesn't see that extra six grand in savings, you are all going to have some explaining to do...

The Ratings Game

Many light fixtures come in many configurations. It is important to look at each fixture as a whole to determine exactly how much power they use, and how much usable light they make.

It's not as simple as it looks, either. Wanna get weird? Modern fluorescent lamps need a "ballast" to start and maintain the light output of the lamp. Why? Fine. You asked for it:

NERDINESS!

Fluorescent lamps produce light when a random electron bounces of an atom of mercury vapor within the lamp. If that electron has enough kinetic energy, it transfers energy to that mercury atom's outer electron, causing that electron to temporarily jump up to a higher energy level.

After the collision, that "excited" electron will emit an ultraviolet photon when it reverts to a lower, and subsequently more stable, energy level.

These photons are not at a frequency visible to the human eye, so to be useful they must be converted into visible light. Ultraviolet photons are absorbed by electrons in the atoms of the lamp's interior fluorescent coating (which are made up of materials called" phosphors.") The UV energy then causes the phosphor's electrons to experience a similar energy jump, (and subsequent drop).

Just like the mercury vapor, the phosphor atoms release a photon when their electrons revert to the more stable level. The chemicals that make up the phosphor are chosen so that these emitted photons are at wavelengths visible to the human eye.

When the light is turned on, the ballast emits electrons sufficient to begin this process, and then maintain the reaction at a stable rate. These electrons collide with and ionize noble gas atoms inside the bulb surrounding the filament to form a plasma by the process of "impact" ionization. Unchecked, this evolves into "avalanche ionization" a condition where the conductivity of the ionized gas rapidly rises, allowing higher currents to flow through the lamp. More current results in more photons being released, and the process continues. Why do we care?

Connected directly to a constant-voltage power supply, a fluorescent lamp would rapidly self-destruct due to the uncontrolled current flow. To prevent this, fluorescent lamps must employ a ballast to regulate the current flow through the lamp.

Aren't you glad you asked? Don't worry if that made little sense to you. You can ignore all of that if you just accept that ballasts are necessary piece of hardware for the correct operation of a fluorescent lamp. And not just fluorescents, either. Most lamps used in commercial settings will need a ballast for the same reasons as a fluorescent lamp. Why does it matter? Why do we care?

Because ballasts use electricity too, and because they are not 100% efficient, they have to be accounted for in the wattage rating of a fixture. Even worse, different ballasts may affect the light output of the fixture as well, and that too has to be accounted for.

So, when your vendor presents his lighting project to you, it is critical to take a close look at exactly how many watts-per-fixture your vendor is claiming the product uses. If you are being presented a fixture with three 50W lamps, you can be very confident that the total watts will be MORE THAN 150W. That tidbit may not find its way into the sales presentation.

Such silliness is not limited to the venerable fluorescent lamp, either. LED lights are incredibly efficient and are the hottest lighting upgrade available at the time of this publication. How does an LED work?

Remember, you asked!

The Ugly Version:

• The LED consists of a chip of semiconducting material doped with impurities to create a p-n junction. As in other diodes, current flows easily from the p-side, or anode, to the n-side, or cathode, but not in the reverse direction. Charge-carriers—electrons and holes—flow into the junction from electrodes with different voltages. When an electron meets a hole, it falls into a lower energy level and releases energy in the form of a photon.

• The wavelength of the light emitted, and thus its color, depends on the band gap energy of the materials forming the p-n junction. In silicon or germanium diodes, the electrons and holes usually recombine by a non-radiative transition, which produces no optical emission, because these are indirect band gap materials. The materials used for the LED have a direct band gap with energies corresponding to near-infrared, visible, or near-ultraviolet light.

• LED development began with infrared and red devices made with gallium arsenide. Advances in materials science have enabled making devices with ever-shorter wavelengths, emitting light in a variety of colors.

• LEDs are usually built on an n-type substrate, with an electrode attached to the p-type layer deposited on its surface. P-type substrates, while less common, occur as well. Many commercial LEDs, especially GaN/InGaN, also use sapphire substrate.

The Quick Version:

When you run some current between layers of weird stuff, they exchange electrons. If you run it backwards, they kick out radiation in the form of UV, infrared, and sometimes visible light.

Even LED's require careful attention to the wattage is considered from the perspective of savings. Remember that the whole shebang (lamp, fixture, ballast/driver) contribute to the overall wattage rating. More critically, the lumen/watt rating. But that's the next section.

Most typical commercial light fixtures will have a wattage rating with your local utility. Ask to see that, and make sure it is a neutral party that calculated the rating.

Light is Light, Except When It's Not...

There are two main ways to quantify the amount of "light" a lamp produces. It all starts with the *candela*,[13] which is the lighting output of one candle (more or less).

1. Commercially, lamps will be rated in lumens. What is a lumen? A lumen is a unit of luminous flux and represents the quantity of visible light produced by a light source. For scale, one normal candle[14] will put off about 12.5 lumens.

2. Foot-candles, typically defined as the illuminance on a one-square foot surface of which there is a

---

[13] *The candela is the luminous intensity, in a given direction, of a source that emits monochromatic radiation of frequency 540×1012 hertz and that has a radiant intensity in that direction of 1/683 watt per steradian.*

[14] *Which gives off about one* candela *of luminous intensity.*

uniformly distributed flux of one lumen. One foot-candle is one lumen per square foot.

These are different things! A lumen is NOT a foot-candle, and a foot-candle is not a lumen! A lumen is an absolute unit of total lighting output, and a foot-candle is the density of light on an object or surface. One is an absolute quantity measured at the source, the other is the visible brightness over a given area at the spot that actually matters. This distinction makes all the difference when applied to your actual space.

Why?

Because with lighting upgrades, it will often be pointed out (ad nauseum) that a "new" fixture or lamp will produce the same lumens as the old for a fraction of the energy. This is important because saving lighting energy isn't cool if you do it by putting in products that produce a lot less light. The vendor needs to demonstrate that he or she is giving you the same quantity of light you had before, because otherwise, you could just turn a few lights off yourself and achieve the same result.

This is fundamentally how you light a space: by spreading light over an area. Lumens are your quantity of light, but how many lumens your lamp produces is not the only factor that determines how well-lit your various areas are in practice. They don't account for the quantity of space being lit.

So, lumens are the light output of the lamp, but the actual illumination of the space is measured in foot-candles. It's foot-candles that determine how well you see the papers on your desk or the stairs in the stairwell. When it comes to foot-candles, not all lamps and fixtures are equal even if they produce the exact same quantity of lumens.

Take a look at the two lamps below.

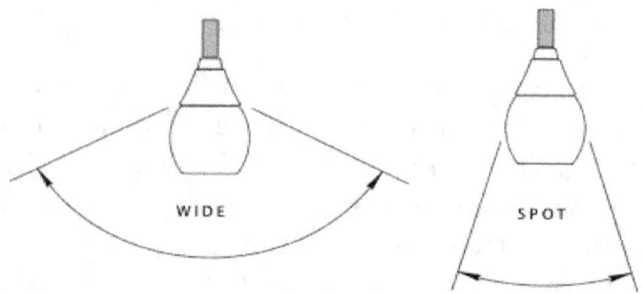

Both produce identical lumens. That is to say they both produce the same quantity of light. However, the lamp on the right puts all the lumens into a smaller area. This increase in light density means more foot-candles, but less area illuminated.

This will light that small area very intensely but leave other areas in deep shadow. If you put this lamp on a sidewalk, you will end up with very bright circles of light with deep dark shadows in between them. Not good.

The lamp on the left will spread its lumens over a larger area, giving fewer lumens per square foot (foot-candles), but covering more square feet of area. As such, it will not light any given area as brightly, but will likely do a much better job of lighting a larger space evenly. This one will keep your sidewalk awash in gentle, safe, helpful illumination.

So, if we assume the lamp on the right makes 100 lumens, and condenses them to one square foot of illuminated area, then it (ideally)[15] produces 100 foot-candles over an area of one square foot. That one square foot area is *illuminated* to 100 foot-candles.

---

[15] *Technically, some of that light will be absorbed by the surface itself, or may have been scattered in transit by dust, etc. Let's keep this simple, OK?*

If the lamp on the left produces the same 100 lumens, but spreads them over ten square feet, then each of those square feet will (ideally) have been illuminated to 10 foot-candles. (100 lumens/10 feet = 10 lumens/square foot or 10 foot-candles.) The larger illuminated area results in less light density. The individual discrete units of area will be less bright, but you will have illuminated more space. More space that you have only illuminated it to $1/10^{th}$ the intensity of the other lamp.

The relationship between how much space you have and how brightly you need it to be lit is how you determine what equipment is and is not a good fit for your facility.

Now let's add a fixture to the mix. Because lights don't float in space magically. You have to mount them to stuff. Fixture geometry plays a large role in how much light gets from the lamp to the floor. A certain quantity of the light from your lamps goes in directions that aren't all that helpful. You want your light to hit the floor, the desks, the display shelves, and all the other areas that are important to you. You do not want your light to bounce around in a fixture or illuminate the recesses or your ceiling. Let's think about our sidewalk again. Consider the following potential lighting connundrum:

You have a parking lot and walkway to your building. The second-shift crew would like to have this area illuminated so they can safely park and enter the building. This seems a reasonable request, and the law pretty much says you have to give it to them. You, being mindful of the mentality of your finance department, go and purchase the least expensive fixtures and lamps available and have them installed. Each of these luminaires[16] puts out 1,000 lumens and takes 100 watts (10 lumens/watt).

You also take some light readings afterwards, because you've read this book and are thus an informed consumer. You find, that on average, your parking lot and walkway are illuminated to 15 footcandles. Not bad. That's actually about right for a parking lot. Your people need sufficient light to avoid tripping, not enough to assemble watches or read fine print.

But then you notice something as you take these measurements. The walls of the builing are lit up as well, as are the the neighboring buildings and half the street and even a goodly chunk of the lawn. Clever individual that you are, you realize that a significant portion of the light is scattered outward and upward. This is not a good thing. You paid good money for those photons and a bunch of them are not hitting anything useful to you. The light from your fixtures and the energy you bought to make that light is being projected out into the air to bounce off of walls and trees and a bunch of other things that you do not necessarily need to be illuminated. You now have foot-candles in places you don't need them. So, a portion of each of the total light output (1,000 lumens) from each luminaire is *illuminating* (putting lumens onto) stuff that you don't need lit up.

As a result, you are buying more electricity to produce a bunch of lumens that don't contribute to the illumination you are trying to achieve on the surface of the parking lot or sidewalk. You decide to call your local vendor and get a free lighting audit. The vendor, being smart, competent and respectable, recommends a luminaire package that

---

[16] *A luminaire a combination of fixture, lamp, and any other accouterments that constitute the whole lighting apparatus.*

directs the light to exactly the areas you want it, and nowhere else.

If this vendor used the same number of lumens, then the smaller area illuminated will be much brighter...as the total light output is the same, while the area illuminated is smaller. You may now have a 30 footcandle average on your parking lot.

Look at your desk right now. There is a good chance it is lit to 30-35 footcandles. Fifty if your office is very bright. Lighting a parking lot to 30 footcandles is entirely unnecessary. Your vendor knows this and recommends a 50 watt lamp, and saves you some energy. That's how the good ones go: The right combination of equipment is employed to achieve the correct level of illumination for the job.

By using a better-designed fixture, you can put more light where you want it, and subsequently get away with a lower-power lamp to get the correct light density (foot-candles) on the surfaces you need it. Spending money on lighting energy that does not specifically achieve the goal is inherently inefficient.

The goal of every light ever produced is to *adequately illuminate a space*. If we hearken back to thermodynamic continuity, it goes like this: It takes electricity to make lumens, and lumens on the target to make foot-candles.

If you can get more foot-candles where you want them with fewer watts than you were before, then you are saving energy. Better lamps produce more lumens per watt, and better fixtures get that light to where you need it with less waste.

So where is the myth? It is the Myth of the Quantifiable. How do you know you are getting enough foot-candles where you need them? Do you know how many you have now? Many vendors will not test light levels unless you ask them to. Typically, they will do what is known as a "lumen-for-lumen" replacement, and make sure you have the exact same lumens as before, but with no real understanding of the actual usable illumination requirements of your space. Your savings will still be great, right up until you have to install a bunch of extra fixtures to keep your building bright enough.

Lighting vendors know how to take light readings and do a good lighting audit. Make them do it and make them promise to match or exceed the illumination (foot-candle) levels you have now.

Degradation, Depreciation, and Boiling the Frog.

There is an old saw that states you can boil a frog alive as long as you do it slowly enough.[17] Your lighting vendor may be trying to do the same to you (metaphorically). How could something so base and cruel be accomplished? By exploiting a little-known facet of commercial lighting products.

Pretty much all lamps will produce light in excess of their rated level for the first few hundred hours of their life. Gradually, they dim to their rated levels and even past them. This happens because lenses get cloudy, dirt and scratches accumulate on reflective surfaces, and other unavoidable factors of aging conspire to reduce the light output of the fixture. This is not a big deal, because it is a known factor and the lumen ratings are based upon the average output, not the initial output. Everybody (in the industry) knows that the lamp is going to over-perform for a little bit before settling in, and appropriate ratings are applied to reflect this.

---

[17] *What sick psycho figured that out?*

Where things get sketchy is when a vendor exploits this to sell you too little lamp for the job. If a vendor is trying to produce really impressive savings numbers, it may become tempting to specify a lamp that produces 10-15% fewer lumens than the current version, just to increase the savings (less lumens means less watts, after all).

When the lights go in they will over-perform (as they do) and initially produce the same light level as before. No one will notice the switcheroo ever happened and the vendor will skip off to the bank with the ill-gotten gains. Over time, the lights dim, but it happens slowly enough that the average person won't notice. By the time you do notice it is too late to get the vendor to fix it.

It is a nasty, dirty trick, and fortunately most vendors won't use it.

Who's Counting?

How many lights are in your building? How many lamps? How many ballasts? What wattage? I bet you don't know, and that's okay. It's not a critical piece of information most of the time. Of course, when the lighting vendor shows up, he or she will be happy to count them for you.

I will be honest with you, dear reader. I have worked for several companies. Some were scrupulously honest and held themselves to the highest possible standards. Some of them clung less ardently to the strictures of uncompromising ethics. No company that I have ever worked for produced flawless lighting audits regularly. You are essentially asking a person who has never been in your building to find every space, count every light, and identify all of them correctly on the first pass. All while simultaneously knowing every quirk of the space and its (often sordid) electrical service history. It is a repetitive, tedious, and mind-numbing task for any building of appreciable size.

I have known auditors who were good enough to do this reliably. Those men and women constituted a fairly small percentage of the auditors out there. I will even go so far as to defend the industry on this one: It's not (always) their fault. The task can be nearly impossible in a big building. As a customer it is in your best interest to help your auditor get it all correct.

A wise building operator will help the auditor find all the hidden rooms, correctly identify the lamps and fixtures, and estimate a reasonable number of run hours for each space.

What happens without your help? The good companies will try their best and often fail, and the bad companies won't care.

This Actually Happened...

A manufacturing facility wanted to update their very old, very inefficient lighting. They were a high-tech machine and assembly shop doing work for aerospace and chemical companies and they desired a very clean high-tech look to their lighting.

My company went through and thoroughly audited the place. Ultimately, we recommended a high-end LED tube lamp to replace their existing fluorescent lamps. The savings were enormous, and everybody jumped on the job.

Upon completion, the installation looked amazing. LED light is incredibly bright and cool, and the results were that this high-tech facility now had a sci-fi lighting effect that made the whole place resemble the bridge of the USS Enterprise[18].

We were pretty proud of ourselves…until we started receiving complaints from the customer that the space was way too dark. This baffled me, as we had allowed for many extra lumens in the design. Furthermore, as we walked through, it was obvious that the space was extremely well-lit. Naturally, we did a complete re-audit and measured foot-candles in all areas. The results demonstrated that we had significantly exceeded their previous illumination levels everywhere and doubled it in a few places. But even in the face of overwhelming scientific evidence, the complaints persisted. Several members of the board were convinced the space was now darker.

When pressed, it was determined that the individuals who felt the space was darker were referring to the area ABOVE the fixtures. You know, thirty feet up in the rafters and where the roof met the wall. They felt that those areas were not as well-lit as they were before.

And they were right. We had designed the lighting to be efficient, and part of that was eliminating all the wasted light getting directed at the ceilings and roof-wall joints. We were a little baffled as to what to do. We had installed what we said we would install, we had properly illuminated the space, and the savings were being realized.

Fortunately, the customer accepted that we had done the job we said we would do and was gracious enough to pay extra to have up-lights installed to restore the brightness of their ceiling space.

The moral of the story is that your version of "wasted" may not be the same as your vendor's! Know what you need and want before they start running wires.

---

18 *The space one, not the aircraft carrier.*

When is it a Good Idea?

Lighting retrofits are the number one selling ECM in the business. This is not a coincidence. They really do generate very good savings at a very good price, typically.

As long as you avoid some of the more egregious sales transgressions, you will likely be very happy with newer more efficient lighting.

If you haven't done a lighting replacement job in the last five years, you should really be looking at one right now. Find a good vendor, get a good audit, and (most importantly) help them design the best project for your facility. The numbers may surprise you!

Myth #2: Your Windows are Bad! (Buy Mine).

Ah, windows!

If there is a more maligned and reviled waster-of-energy out there, I do not know what it is. Architects love 'em. Engineers hate 'em. You can certainly rest assured that everyone out there has a better version to sell you. These are big-dollar items and many vendors will tell you that they are big-time energy-savers at the same time. But are they really?

We all know that leaky windows, or single-pane windows, or old windows are total energy hogs. Windows are basically giant holes in your building sealed with a thin membrane of one of the worst insulating materials ever made. Many vendors will point this out at length and then regale you with the wonders of modern double, triple, quadruple or whatever-pane windows. Often these super-windows are filled with exotic gasses and the frames are constructed with space-age materials. All of this contributes to a modern window that is many times superior to the old-school version currently blighting your building.

And for the most part, modern windows ARE many times better than the old ones. That's just science at work. So, what's the problem? The problem really boils down to the old carnie trick of misdirection. The vendor is not lying when they make lofty claims. They are simply leaving out one very important piece of context.

Problem #1: Much Better than Awful is still Pretty Bad (the Myth of Magnitude)

Heat is gained or lost through a building's exterior via two methods: Infiltration and Conduction.

Infiltration is warm air leaking either into or out of the building, and conduction is the transfer of heat through the physical materials of that construction. The two formulas for this are:

Conductive heat loss: $U * A * \Delta T$

Infiltration heat loss: $1.08 * CFM * \Delta T$

Where:
U= U-value of exterior construction
A= Area of exterior surfaces
CFM = Cubic feet per minute of outside air
ΔT = Difference in temperature between indoor temperature and outdoor temperature.

Energy wasted through windows will follow these formulas. Glass will lose heat the same way any other material will; and thus, the same formulas apply. Namely:

$U*A*\Delta T$

U is the rate of heat conduction through the material and is very important to why new windows often don't save as much as you would think. A standard sheet of window glass has a U-factor of 1.16 (ish). We can go through a lot material science and math to demonstrate exactly what that means, or you can take my word for it that any material with a U-value of 1.16 is an objectively terrible insulator. However, we knew that much when we started. That's why windows bleed so much stinking energy from a building.

A modern, double-pane, argon-filled window will have a U-value of about .42. That makes it 2.4 times as good at insulating as a standard single pane window. That's called "R-Value." R-Value can be considered the inverse of U-factor, and pretty much measures the exact same thing. It is considered a measure of the materials "resistance" to heat transfer. By using the inverse of U-factor, we get nice, easy to understand numbers that increase as the quality of the insulator improves. This makes it easier to market to non-technical consumers. Explaining that something with an R-value of 10 is better than an R-value of 5 is easy. Explaining that U-factor of .1 is not as good as a U-factor of .05 tends to lose people and is less impressive on the packaging.

At this point we are all going along for the ride. a U-factor of .42 (or an R-value of 2.4...same thing) is much better than a U-factor of 1 (Which is also an R-value of 1! And you thought you'd never use algebra again!). That is a HUGE improvement. Now let's go ahead and put it in the context of an actual building to see where it all goes so horribly wrong.

Your building wall likely has a U-value of .06 or so. That's seven times as good as your fancy new windows. If you have 1000 square feet of window at U=1, and your average ΔT is 55 degrees, then your old crummy window will lose 55,000 BTU/hr.

$$U*A*\Delta T = 1 * 1000 * 55 = 55,000$$

Through a tough New England winter this could easily cost you $1000-$2000 in natural gas and a lot more if you are on electric heat. Our new, U.42 windows will only lose about 23,100 BTU/hr. under identical conditions and save you as much as 1,200 bucks.

$$U*A*\Delta T = .42 * 1000 * 55 = 23,100$$

But what about a *really* awesome window? What about a triple-pane, argon-filled uber-window? We are talking about a best-case U-factor of .2. That is actually pretty darned impressive! Your best-case savings in the above example have gone from $1,200 to the princely sum of $1,600.

1,200-1,600 dollars a year ain't terrible. That's real money. But let's take our engineer hats off and put our accountant hats on. I will not attempt to estimate what the installation of 1000 sq. ft. of exotic new windows will cost, but I think we can all concede that the answer will be "many thousands of dollars." Generally speaking, the simple payback for replacing windows in a commercial building is almost always 25 or more years. Not because you aren't saving energy, but because windows are monstrously expensive.

Keep in mind, this case assumes a very bad winter. Furthermore, I can assure you that you do not want to know what a triple-paned, argon filled window costs to install, and even worse, replace if it gets broken.

Leaks (infiltration) fall into the same sort of trap. Your new windows, with professional installation, are likely to be leak-free. But let me ask you this: As a building operator, what is the most cost-effective solution to your leaky windows? If you answered, "caulk and weather-stripping" then go ahead and pat yourself on the back for being clever. If you answered, "tens of thousands of dollars in new windows and a lengthy and disruptive installation," then I can assume you probably work for the government.

The take-away here is that new windows will absolutely save energy. That is the very essence of the Myth of Magnitude. If compared to old windows, the energy saved seems very significant. If compared to the costs of new windows, they become just about the least cost-effective ECM you could ever attempt. This does not mean that you should never replace old windows. There will come a time when you will have to address the windows on your building, but energy savings are not going to be a viable method of off-setting those costs.

This Actually Happened...

Working as an Energy Engineer for a large Energy Services Company (ESCO), I was tasked with evaluating the claims of a window manufacturer that was interested in supplying us with products for retrofits.

Most of these projects were municipal in nature, and the governmental regulations pertaining to performance and savings modeling (calculations) were stringent and numerous.

The work was plentiful and profitable, and as a company we were resigned to dealing with the massive and brutal oversight required and soldiered on. Enter the window vendor.

This vendor had a very impressive window product that purported to have an R-value of 6 (U-factor of .17), making it orders of magnitude better than any other window we had encountered. Which was a good thing, since they were more than twice the price of a regular high-efficiency window.

The vendor's initial presentation was obviously intended to be a sales presentation and administering it as such to a room full of engineers was a failure of epic dimension. Upon hearing requests from us to quantify the savings potential and justify mathematically the very high claimed potential of these windows, the presenter was incapable of supporting any of the claims with documentation stronger than "our engineers are very smart," "our tests have shown" and "our models demonstrate that…"[19]

This was not going to be adequate for our governmental overlords, and with the grit of true corporate soldiers we pursued quantitative analysis of the product far harder than was prudent. The inventor of the window, the engineers who built the window, and the executives who ran the window company, were without exception incapable or unwilling to produce any test data or third-party analysis for their product. Furthermore, they were disinclined to allow us to analyze it.

Of course, because this was a large company and they had a huge high-profile project to exploit, the windows were used anyway. To give proper credit where it is due, the windows performed fairly well. They did *not* perform better than a regular modern window at half the cost would have, however.

When are New Windows a Good Idea?

This is a case where the obvious answer is the right answer. The best time to by energy efficient windows is when you need new windows. Like so many parts of your building's exterior, windows are a sacrificial fixture that will need cyclical maintenance and eventual replacement. Eventually the day will come that your windows will need to be replaced. That is the time to go and buy the best window you can afford.

Windows are energy hogs because they are holes in the building. Big gaping holes that let cold air in and heating energy out. We cover them with a thin shield of glass to keep the breeze out, and having the best possible shield is always a good idea. But realistically speaking, until a

---

[19] Argumentum ad vericundium!

magical transparent insulator is invented the savings from exotic modern windows will never come close to covering their costs. This is an ECM best attempted when the windows are due for replacement anyway. At which point only the incremental increase in costs for the more modern technology needs to be compared to the increased savings over just replacing with the same inferior windows. The "incremental cost/incremental benefit" ratio is far less of an obstacle to installation.

Myth #3: Tinted Windows-Not Just for Cars Anymore!

Interestingly, there is another common method for improving the performance of your windows that does not cost nearly as much as replacing them. There are multiple vendors purveying window tinting and window film that purports to save significant energy by preventing solar heat from getting in, and interior heat from getting out.

It starts with a solid premise. Sunlight comes in through your window and brings with it energy. No real surprise here because sunlight *is* energy. This energy passes through the windows of your house, strikes the solid objects within, and heats them up. There is nothing revolutionary in this assertion either. If you have ever stood in a sunbeam, you have experienced this effect. When this happens to your building in the summer, it adds extra load to your air-conditioning system. This is referred to as "solar gain." We remain safely within the realm of good science.

Even worse, in the winter time the infrared radiation from the heated surfaces and warm air in your space will subsequently escape by heating the window glass as well. This will increase the load on your heating system.

If we could somehow prevent that infrared emission and infiltration, you can keep the good rays in and the bad ones out. Low-emissivity window films purport to do this very thing.

Problem #1: Your Window Does Not Know What Season it is.

If I started to lose you up there, then pat yourself on the back. There is a lot wrong with this narrative.

Sunlight coming in through the window absolutely adds heat to the space. In the summer this means bigger cooling loads. Of course, in the winter time that extra solar gain is mighty helpful. It actually heats the space a little bit and takes some of the load off the boiler. Does the window film know when winter has arrived, and subsequently allow the sunlight in when it is needed?

No. No it does not. So, all those sunlight-borne BTUs that were helping you in the winter time are gone. Sure, you may have relieved some of the load on your cooling system, but in any heating-driven climate, the loss of the solar gains in the winter time completely offsets the cooling kWh saved in the summer.

Why is that? Above the Mason-Dixon Line, the heating season is often twice as many months as the cooling season, and it is a 24-hour a day requirement. Up here, heating a building is typically far more expensive than cooling it. Saving a little bit of cooling energy in the summer at the expense of adding a lot of heating energy in the winter makes no financial sense.

Many window film vendors will attempt to brush the problem aside by pointing out that their product prevents the infrared inside the space from getting out, as well.

Well, which is it then? Is the film keeping heat in or letting it out? Does it know when the window is in heating season and cooling system? What magical properties of the tint tell it when to let infrared energy in and when not to? On to the next section...

Problem #2: You Just Ain't Leaking That Much Infrared.

When you add heat to your space, you don't just heat the air, you heat everything in it, too. The file cabinets, the desks, and that stupid poster of the cat that says "Hang in

There[20]" all get heated to room temperature (or thereabouts). Even the window glass itself experiences a rise in temperature. There's a lot of sunlight hitting it after all.

Since everything above absolute zero gives off some thermal radiation in the form of infrared, it follows that all of that stuff is giving off infrared, too. If your windows do not employ window film, then the glass itself (having been heated by either the space or the sun) will then emit that infrared (from inside) to the outside, because one whole side of the window is completely exposed to the outdoor air.

Window film vendors will advertise that they reduce the "emissivity" of the glass so that less of that infrared will be lost in this manner.

Of course, the sun beams infrared (along with every other wavelength of radiation) to the earth all day long. It is safe to say that there is a big difference between the omnidirectional, weak, and diffuse radiation of a 70°F pane of window glass, and a giant, self-sustaining fusion reactor 865,374 miles across delivering 1,368 watts per square meter of pure energy (only 50% of which is infrared) all the time. One might call that difference "astronomical."

Even if your window is shaded all day long, more infrared from the sun will hit it on a cloudy day than will ever escape from inside. To say that a one-micron-thick sheet of plastic can prevent window glass from emitting infrared both inside and outside beggars belief.

It all boils down to the temperature of the pane of glass. In the summer, the sun will want to heat the glass above the room setpoint, creating load on the cooling system. In the winter the room will want to heat the glass above the

---

[20] *What is this? 1981? Get rid of it, man.*

outside air temperature, placing load on the heating system. The film can only reduce emissivity (how quickly the glass radiates heat in the form of infrared) a tiny bit and it cannot control when it does so. What you gain in summer, you must lose in winter.

This Actually Happened...

My first job in this industry was as a field survey technician for a consulting firm. This firm was run by a vicious old codger[21,22] who took everything very seriously, all the time. Even the little stuff that no one else cared about, which it turns out, was kind of important to this story.

The company was hired to be a consultant and owner's representative to a municipality having multiple buildings retrofitted. When evaluating ECMs for the school district, one of the measures in question was several hundred thousand square feet of window tint spread over multiple buildings. Of course, the window tint was only a tiny part of a massive project and nobody cared about it at all. It was an afterthought.

The company owner, after a stern evaluation of my skills, determined that afterthoughts were exactly as much responsibility as I could handle. Subsequently, he gave me the enviable and high-honor task of counting every god-forsaken window in the district and calculating the square footage.

So I did.

Imagine my horror when it became apparent that the ESCO in question had sold 400,000 square feet of window tint to a district that only had 250,000 square feet of window.

---

[21] *Hi, Dad!*

[22] *I take offense at being called "old"!-RRV*

Nervous and lacking confidence, I counted the whole thing again, certain that I had made a mistake. I had not.

I told the boss my findings, and we go and count them again just to be sure. The count was definitely good, and it was obvious that the window tint had been significantly oversold. At the next progress meeting we confronted the ESCO and they insisted that we must have missed some windows. Fine then, we all went out together and counted them again. For the fourth time. It was still a good count.

Finally, the tint vendor was brought in and guess what? He was *sure* we had counted wrong. I was no longer youthfully insecure, folks. No, at this point I was a window-counting maestro with skills homed in the hard winter world of exterior fenestration enumeration.

Oh? You wanna count again? Let's do this. Except now we are all going together: Vendor, ESCO, town facilities manager and anyone else who wants to come along. It was about nine people in all.

This time, I made the vendor admit to the town and the ESCO at every building that my counts were accurate and his were not. Verbally, every time. If I tell you that I took no satisfaction in this, I trust that you will take my word as a gentleman on it and not cast aspersions[23]. Sure enough, after counting five times[24], and dragging many people out for many days, the vendor admitted that they had only "estimated" the square footage. The vendor had to agree to tint virtually every other building in the town to make up the difference.

------

[23] *Felt awesome.*

[24] *To complete the picture of "vicious codger" I must point out that I only did it once. After that he was on his own. -RRV*

So, if you are going to buy window tint, count your own windows. It seems silly but fibbing on the square footage is the easiest way to pad the bill[25].

When Does Window Film Make Sense?

Window films absolutely save cooling energy. They reduce solar gain and that means less heat from the sun gets inside the building. If you do a lot of cooling and very little heating, then this is a good thing.

If my building was in Florida, or Hawaii, or Arizona, I would absolutely recommend window tints and films. If you want to keep heat out they absolutely help. I would not get too excited either way. The bulk of the energy lost through a window is via conduction (the contact between the glass and the outside air), not radiation.

If you have more than 1,000 heating hours annually, this measure will start to hurt more than it helps.

---

[25] *Dumb vendor! Everything else was "stipulated". Counting the square feet was the only thing that could be verified. -RRV*

Myth #4: Turn Down Your Domestic Hot Water!

For our next myth we are going to do a little algebra. Consider the energy savings claims from reducing the temperature of domestic hot water. This is a standard recommendation in almost every energy conservation manual. It's simple, intuitive, and almost HAS to save energy, right? Let's just explore for a little bit exactly *how* this will reduce the energy consumption and how to calculate the savings.

Problem #1: You Haven't Really Changed Anything (Myth of the Intuitive)

In one state's official Energy Conservation Workbook for Multifamily Housing, measure number 301 states:

> *"Reduce water temperature: For every 10° F that water temperature is reduced, approximately 1.7 million Btu's will be saved per apartment per year."*

Who knows where that number came from? (By reverse engineering, 1.7 MMBtu at a 10° F reduction means that 56 gallons per day was the base number.)

In another state's Energy Workbook for Office Buildings, conservation measure number F1 states:

> *"Hot water should be stored and distributed at the lowest useful temperature.... Normally, 110° F water is adequate for domestic use."*

They go on to give the following formula to calculate the savings:

Q= Annual gallons X 8.35 X (T-110)

I'll skip the derivation and tell you this equation is mathematically correct. But the equation does not specifically or accurately address the domestic hot water system and use.

Suppose that 100° is the desired temperature at the faucet. This will ultimately be determined by the hands of the user. If the water temperature is 140° the user will add cold water to achieve the desired 100°.

Let's assume that the faucet in question delivers two gallons per minute of water. If the hot water is 140° and the cold water is 60°, then to achieve 100° (on your skin), the ratio of hot to cold must be 50%/50%. If 50% of that two gpm is hot water, then the consumption of 140° water will be one gpm. (Because 50% of two is one. Math is fun!).

Since your water comes in on a pipe at around 60°F, then your heater has to raise it from 60° to 140° to do its job. One gpm at an 80° F ΔT (60° heated to 140°) will consume 40 MBh.

Lowering the hot water temperature to 110° will cause a change in the hot to cold ratio from 50%/50% to 80%/20%. If 80% of the two gpm is hot water, then the consumption of 110° water will be 1.6 gpm. This smaller ΔT, when multiplied by the (now larger) flow rate of 1.6 gpm, will yield the same 40 MBh.

50° F ΔT (60° heated to 110o) X 1.6 gpm = 40 MBh

There will be no energy savings from the temperature reduction. Completely counter-intuitive, but physics does not really care about your intuition.

Problem #2: Too Much Water, Too Much Variation, and the Myth of the Quantifiable

The calculation method in the first manual predicts annual savings of 5.1 MMBtu per apartment. The other manual comes up with the same number if you allow 56 gal/day of HOT water for 365 days per year or about 28gal/person/day. Depending upon how many people are using the space, one could consider that a very large quantity of domestic hot water. Both of your humble authors grew up in a house that had anywhere from six to fourteen people living it. This still seems very high for an average. The research seems to indicate that the averages will be much lower for most commercial buildings.

Type of building	Consumption per occupant		Peak demand per occupant		Storage per occupant	
	liter/day	gal/day	liter/hr.	gal/hr.	liter	gal
Factories (no process)	22 – 45	10	9	2	5	1
Hospitals, general	160	35	30	7	27	6
Hospitals, mental	110	25	22	5	27	6
Hostels	90	20	45	10	30	7
Hotels	90 – 160	20 – 35	45	10	30	7
Houses	90 – 160	20 – 35	45	10	30	7
Offices	22	5	9	2	5	1
Schools, boarding	115	25	20	4	25	5
Schools, day	15	3	9	2	5	1

This table covers total potable water consumption. The actual portion of this water that is "hot water" will vary greatly.

Certainly, the heat loss in the distribution piping and storage tank will be lower if the water temperature is lower. That is seldom calculated because it is usually considered insignificant (or too difficult.) The Myth of the Quantifiable is strong in this one.

This measure is sort of the most basic "Myth of the Intuitive" that one is likely to see. On the surface, this seems a fairly straightforward measure. But it is not, and the potential for misapplication is huge.

When is it a good Idea?

Almost never. If your hot water tank is set at a reasonable temperature, and your needs are being met, the only thing turning it down will do is make it not meet your needs.[26] If you want to save DHW energy, set the temperature somewhere reasonable and insulate the tank and all the piping.

---

[26] *Which actually will save energy until all the complaints make you turn it back up again.*

Myth #5: Fun with Inrush Current

Here is a fun fact: When a motor is started, it can draw from 300% to 600% of its nameplate current for a few (typically 1 – 3) seconds.

This is sometimes called "surge current" and is caused by (many boring things that go on inside a motor that we will talk about later). Many people call it "inrush" current because that is actually a darn good way to describe what is happening: When you throw that switch a massive swooshing tidal wave of electrons gets pulled into the motor. The inrush current has often been identified as a cause of high peak electrical demand charges.

Peak electrical demand is measured by utility companies and billed separately to accommodate the capital investments required to ensure that all customers will receive all the electricity that they want at any time. The two most common measurement periods are fifteen and thirty minutes. The highest electrical rate of consumption (demand) for a fifteen-minute period is charged for the entire month (sometimes for eleven months if there is a ratchet charge). It is economically important to keep this value as low as possible.

Often, it is recommended that buildings with many motors to do one of two things:

1. Stagger start: Stagger motor start times to keep all of that current from driving the demand meter up at once and crushing the poor customer with demand charges. Good news! They will sell you stuff to do that!
2. Install soft-starters to slowly bring motor current up and avoid large inrush loads. These are not free.

Problem #1: Tiny Time Units and the "Myth of the Intuitive"

There is no nice way to put this, so let's just air it out. Because of the way that the peak demand is measured by electric utility meters the inrush current has an insignificant impact on the measured peak electrical demand for a facility. Therefore, taking pains to stagger the start of electrical motors simply to minimize the combined impact on demand charges is like trying to turn the lights off and jump into bed before the room gets dark.

But that makes no sense! If ten 100kW motors kick on at 300% inrush, doesn't that equal:

100kW*300%*10 = (3000kW)?

That's a huge number! What's the deal?

The answer is in the fine print. People often state that the peak demand is the highest fifteen-minute demand for the month. These people are wrong. The correct statement is that it is the highest fifteen-minute *average* demand for the month. The demand is measured by accumulating kWh (energy consumption) over a fifteen-minute period and then dividing by the number of hours (0.25 hours) to obtain the *average demand*. For example: 100 kWh accumulated in a fifteen-minute period goes like this:

100kWh/.25 hours = 400kW

So, the real question is: how much does inrush current affect the accumulated kWh? Math time!

Assuming the following motor parameters (related directly in kW to keep things simple):	
Motor nameplate load:	50 kW
Motor inrush:	300% or 150 kW
Extra kW:	100 kW
Inrush duration:	3 seconds

There certainly appears to be a significant demand increase during motor startup. An extra 100 kW is drawn from the system whenever this motor starts. But then again, it only lasts for three seconds! If we apply the formula as we now understand it, the additional kWh ends up extremely small.

3 seconds = 0.0083 hours

100 kW X 0.0083 hours = 0.83 kWh

Therefore, the inrush current will increase the accumulated 15-minute kWh from 100 to 100.83 kWh. Ergo, the effect on the peak demand is:

100.83kWh/.25 hours = 403.3 kW

The increase in demand directly attributable to the inrush current is less than 1% (0.83%).[27]

When Does It Actually Work?

When does staggered starting equipment make sense? Simply put: If you have multiple pieces of equipment that will operate at full load during a lengthy startup period and then "coast" back to a lower operating load. For instance, multistage air conditioning units.

Suppose that you had three two-stage AC units with 25-kW/stage. Arithmetic tells us that the total unit load is 50 kW. During morning cool-down, the AC unit will operate at full load (50-kW) until the space is satisfied. Once the space is at occupied temperatures, the second stage will shut off.

---

[27] NOTE: if the demand period is 30-minutes, the affect is even less.

Therefore, starting one unit at a time, then waiting for it to drop back to single-stage operation will reduce the peak demand caused by starting these units.

Startup without staggering

	#1	#2	#3	Total
1st 15-min	50	50	50	**150**
2nd 15-min	25	25	25	75
3rd 15-min	25	25	25	75
4th 15-min	25	25	25	75

Startup with staggering

	#1	#2	#3	Total
1st 15-min	50	0	0	50
2nd 15-min	25	50	0	75
3rd 15-min	25	25	50	**100**
4th 15-min	25	25	25	75

Myth #6: Everybody Loves Variable Frequency Drives!

Induction motors rotate at a speed proportional to the frequency of the incoming power. Why? Because Nikola Tesla was brilliant[28] and that's all you need to know to cover this myth. A Variable Frequency Drive (VFD) is essentially a motor controller that accepts incoming AC Current at 60 cycles/sec (Hz) and changes that frequency to alter the speed of the motor. If you send power at 30 Hz, which is 50% of the original 60Hz, then the motor will operate at 50% of its nominal speed and much less power.

VFDs are possibly the most ubiquitous and popular ECM sold after lighting. It's easy to see why, too. Much of the time these things save a lot of energy. So why are they in this book? Because we have gone just the teeniest bit crazy with them and it's starting to get out of hand. Not every motor will benefit from the installation of a VFD and much of the time the savings are not as impressive as they might appear at first.

Problem #1: "V" is for "Variable" (the Myth of the Intuitive)

The V in "VFD" stands for "variable" and that is important. A VFD will vary the speed of a motor based upon some sort of feedback from the system it is attached to. Like a thermostat controlling the amount of hot water a coil receives, a VFD will control how much motor energy a system receives. But critically, it needs a *reason* to change

---

[28] *He was also crazy, naïve, and OCD, but nobody's perfect. -RRV*

that speed. Many vendors lose track of this in their zeal to save energy. Here's why:

Motor horsepower (or kW if you are outside of the colonies) is governed by the affinity laws:

Affinity Laws:

If the impeller (pump) or fan diameter is constant, then the affinity laws are as follows:

Volume Capacity (GPM or CFM, typically)
$q_1 / q_2 = (n_1 / n_2)$
Head or Pressure (In. H2O or ft., typically)
$p_1 / p_2 = (n_1 / n_2)^2$
Power (HP, kW, typically)
$P_1 / P_2 = (n_1 / n_2)^3$

Where:
q = volume flow capacity (m3/s, gpm, cfm,)
n = wheel velocity - revolution per minute - (rpm)
p = pressure
P = power

This makes many engineers and vendors giddy. When we consider kW as a function of speed, and the affinity laws do their magic, you get a table that looks like this when applied to a (fully-loaded) 10HP motor:

(Turn the page...)

kW	% Speed
7.46	100%
6.40	95%
5.44	90%
4.58	85%
3.82	80%
3.15	75%
2.56	70%
2.05	65%
1.61	60%
1.24	55%
0.93	50%
0.68	45%
0.48	40%
0.32	35%
0.20	30%
0.12	25%
0.06	20%
0.03	15%
0.01	10%
REALLY TINY NUMBER	5%

Seriously? Cutting the speed by 25% really means reducing the kW by 50%? It sure does, and that is real savings. So let's go ahead and put one on my 10HP air handler fan!

And then what? How do you tell the VFD to change the speed? Unless this is a Variable Air Volume system (VAV), then this system will not experience any *variation*. Otherwise it is called a Constant Volume (CV) system and it is what many buildings employ. They are cheaper, easier to install, and generally do an OK job of maintaining occupant comfort. What it is critical to be aware of with respect to a VFD application is that CV systems pull the same amount of air no matter what. They are sized, designed, and installed to have the same flow through them *without variation*. The "C" in CV is for "constant." The "V" in VFD is for "variable." This is not a coincidence. If there is no variation in airflow then there is no reason to change the fan speed and therefore no reason to change the motor speed. What is often forgotten by vendors is that if nothing is variable, then there can be no savings from a VFD. Worse, you can do real damage if you try to vary it anyway (and people do).

In order for a typical CV air handler to work and do its job correctly it needs constant airflow. The burner, the heat exchanger, the heating and cooling coils, all need certain flows to operate correctly (unless they have been designed for variable flow, which means it's not a CV unit. Tautologies are fun!).

In many pumping systems, constant flow protects the pump from cavitation and dead-heading under various heating or cooling conditions and balance is achieved through the liberal use of 3-way valves.

These valves direct water either through a coil or around it to make sure that even when a piece of equipment is not calling for heating/cooling the flow rate seen by the pump itself never changes. Since the flow rate of the water is constant, installing a VFD on such a system is the same as installing a VFD on a CV air handler fan: no savings.

Consider the control valve on the unit ventilator pictured:

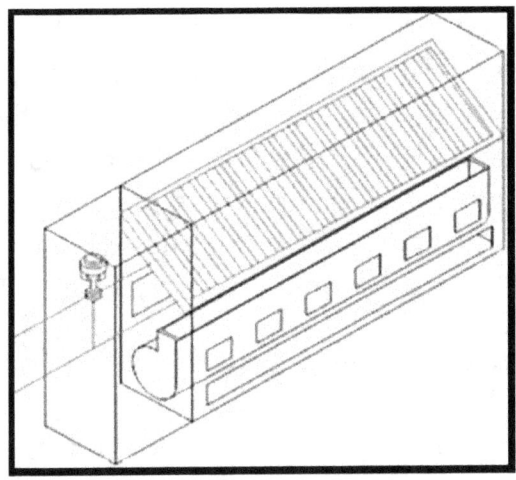

This unit ventilator cutaway shows the 3-way valve and hot water coil. The position of the valve determines when hot water is either directed through the coil or is bypassed to the return side of the loop. The thermostat determines when the coil receives water and when it is bypassed based on space temperature. The volume of hot water itself will not change, merely the route it takes.

If you aren't careful with how you apply VFD technology to your air handling or pumping systems, you can break stuff, too. That will likely impact the financial viability of your project as well!

For instance:
- Insufficient airflow can cause overheating of a heat exchanger in a gas-fired unit, and now you have burned your building down.
- You may starve terminal units and reheat coils further down the system, and compound all your problems again.

- On Constant-Volume Pumping systems, you can starve coils, get insufficient heating/cooling, and ultimately cause cavitation and destroy the pump.
- On a chilled water system, you can reduce the flow enough to get insufficient turbulence in the tubes and ultimately cause the chilled water to freeze, burst the tubes, and destroy the chiller.

Problem #2: Performance Under Pressure and the Myth of Magnitude

Remember that kW table? Let's add another key operational parameter to it: Pressure (p).

kW	p	% Speed
7.46	1.00	100%
6.40	0.90	95%
5.44	0.81	90%
4.58	0.72	85%
3.82	0.64	80%
3.15	0.56	75%
2.56	0.49	70%
2.05	0.42	65%
1.61	0.36	60%
1.24	0.30	55%
0.93	0.25	50%
0.68	0.20	45%
0.48	0.16	40%
0.32	0.12	35%
0.20	0.09	30%
0.12	0.06	25%
0.06	0.04	20%
0.03	0.02	15%
0.01	0.01	10%
0.00	0.00	5%

Whether you are pumping water or moving air, pressure is very important, and pressure is irrevocably connected to volume and speed via our much-beloved affinity laws:

Affinity Law for pressure:

$$p_1 / p_2 = (n_1 / n_2)^2$$

Pressure will vary with the square of the speed proportion. If we reduce speed (n_1)) to 90% (n_2), the new pressure ($p2$) becomes:

$$p2 = p1 * .(.9^2) = .81 \text{ or:}$$

81% of p_1

Pressure, flow, and power are all interconnected via the affinity laws, and you cannot mess with one without messing with the other. Let's assume our 10HP fan makes 1" static pressure and apply the affinity laws again:

Look at that "p" column. That number shrinks pretty fast. Why do we care? What do you think happens when your system has .75" of pressure drop and you reduce the speed to 80%?

kW	p	% Speed
3.82	0.64	80%

Whoops. Your fan is only making .64" of static pressure now. Why is that bad? Because it means that the pressure produced by the fan is less than the pressure drop caused by the ductwork itself.

Essentially, your fan must make enough pressure to overcome the pressure drop of your distribution system or the air stops moving. Even more annoyingly, because flow and pressure are mathematically related, as your fan pressure drops so does your flow. Which is to say that your fan needs to make enough pressure to NOT ONLY overcome the pressure drop of the system, but it must make enough EXTRA pressure to achieve the FLOW you want as well. When you are only producing a little bit more pressure than the resistance of the system, you will only get a little bit of air movement (flow). When you produce a lot of extra pressure you get a lot of movement.

If your VFD reduces the static pressure too much you get too little (or no) airflow. Congratulations, you have starved your space for airflow. But hey! You are saving energy!

So, pressure counts, got it. Now what was that about flow? Flow matters too. Let's look at the table with *all* the parameters:

kW	CFM (flow)	p	% Speed
7.46	12000	1.00	100%
6.40	11400	0.90	95%
5.44	10800	0.81	90%
4.58	10200	0.72	85%
3.82	9600	0.64	80%
3.15	9000	0.56	75%
2.56	8400	0.49	70%
2.05	7800	0.42	65%
1.61	7200	0.36	60%
1.24	6600	0.30	55%
0.93	6000	0.25	50%
0.68	5400	0.20	45%
0.48	4800	0.16	40%
0.32	4200	0.12	35%
0.20	3600	0.09	30%
0.12	3000	0.06	25%
0.06	2400	0.04	20%
0.03	1800	0.02	15%
0.01	1200	0.01	10%
0.00	600	0.00	5%

At 80% speed, your fan is making .64" total pressure and 9600 CFM. If your system needs more than .64" total pressure and/or more than 9600 CFM to operate correctly, then at 80% speed you have gone and broken it. There is no prize for that. Every building system has minimum pressure and flow requirements that dictate how the equipment can and can't perform. These requirements dictate the speeds at which the motor can be allowed to run before performance suffers or fails altogether.

Many vendors will make VFD savings claims based upon unrealistic expectations for what the minimum speed for the motor can be. Building operators and engineers need to have at least a basic grasp of what pressures and flow rates the system requires to operate correctly and ensure that savings predictions account for the minimum speed requirements.

Problem #3: Can't Save kW that Doesn't Exist (The Myth of the Quantifiable)

Remember that boring motor stuff we skipped in the Inrush Current section? Well, here it is. Don't say I didn't warn you:

Nerdy Electrical/Mechanical Stuff Here!

Modern induction motors are funny things. Well. Not really. They're just motors, and as such aren't particularly humorous. But how they operate is a little weird. They have a rotor, which rotates (imagine that!), and a stator, that doesn't. The fundamental concept is that when current is sent through the stator and thus *induced* in the rotor, the magnetic field rotates around the stator and drags the rotor around with it. This is what makes the shaft spin.

Now, this will get hairy but stay with me: the frequency of the induced current in the rotor is determined by the slip frequency/proportion between the rotating field and the rotor speed. Think of it like this: Just because the field is rotating at a given speed doesn't mean the rotor is. Depending how much

---

force (load) is acting against that rotation, the shaft will resist rotating. The amount of difference between the two is called "slip." As the mechanical load increases, the rotor attempts to slow down, which increases the slip proportion *and thus the amount of current induced in the rotor.*

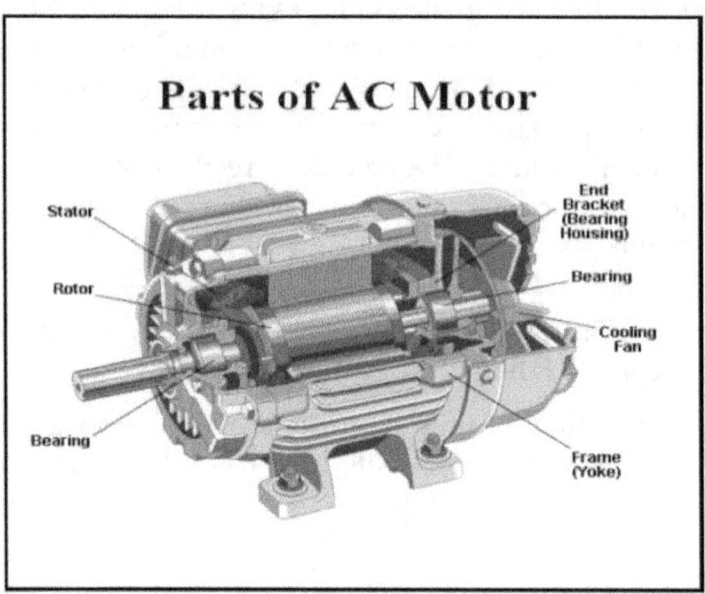

Parts of AC Motor

When electricity flows through the stator, the magnetic field drags the rotor around with it.

Since stator and rotor currents are inextricably intertwined (magnetism!), the stator current is also increased as the load on the motor increases. Therefore, the stator current is always proportional to the mechanical load. This feedback mechanism keeps the motor at or near synchronous speed until such time as the load exceeds the circuit's ability to maintain speed/current, and stuff breaks. This is exactly why inrush current is so big. For a few seconds right after throwing the switch, overcoming inertia represents a huge amount of slip. But the rotor quickly catches up and your current draw stabilizes.

If that made no sense to you, it's OK. The important thing to understand is that *motors only pull as much current as they need to do the job.* This is a problem for a VFD vendor, because if a motor is lightly loaded, it is not drawing much current. If you aren't drawing a lot of current, you aren't using a lot of power. If you aren't using a lot of power, he can't save you a lot money with his pricey widget.

Without clamping a meter on your motor and measuring the current, you simply don't know how much juice it's pulling, and thus you cannot know what potential savings a VFD might get you. Most design engineers will oversize a motor for the sake of safety, which is good design practice. Why? Because "errors and omissions" insurance is expensive.

An oversized motor is not nearly as big a problem as an undersized motor for reasons you now understand. It's only going to pull as much current as it needs, so who cared how big it is? If it was your engineering license on the line you would specify a bigger motor, too. Don't judge.

Many VFD vendors will assume a motor is nominally loaded (running at 100% of rated capacity). This is almost always way too big an assumption. Big assumptions are advantageous to the vendor because they create an artificially high starting point for the comparison between before and after the VFD installation.[29] If they start with an artificially high load and subsequently inflated power consumption, and then compound that with an impossibly low minimum speed (previous section), they can generate the kind of savings numbers that dreams are made of. Furthermore, they will have done so in a manner that is at least mathematically compelling. "Mathematically

---

[29] *"Inflating the baseline" rears its head!*

compelling" is not the same as "practical" or "realistic," of course. But hey man, hammers and nails, right?

Fortunately, this is an easily surmountable problem in that you can simply have a competent professional come and actually slap a meter on your motor and read the amps at load. Now you will know what the potential is and cannot get fooled.

Let's imagine you have a 50HP pump motor that is a good candidate for a VFD. It runs all 8,760 annual hours and produces twice as much flow as you need. You know that a VFD can deliver correct flow for a fraction of the horsepower, so you call your vendor and ask for a proposal. The vendor presents you with a savings analysis that looks like this:

Existing		Proposed	
HP	50	HP	50
kW	37.3	kW	6.25
Hours	8760	Hours	8760
kWh	326748	kWh	54750
$/kWh	0.17		
Existing $	$ 55,547.16		
Proposed $	$ 9,307.50		
Savings	$ 46,239.66		

Forty-six grand in savings? WOW! Sign me up! Since a 50HP VFD and controls probably costs less than $40,000 with installation, we are solidly into the territory of "no-brainer" projects. This would be an amazing savings and an amazing coup for your facility. Your promotion is virtually guaranteed...

At this point, you realize that I am about to dash this beautiful dream mercilessly upon the rocks of cold unflinching reality. If it helps, please understand that I take no joy in it.[30]

This vendor's analysis takes a few liberties. It assumes that the motor is 100% loaded and that your system functions nominally at 50% speed, which will deliver only 25% of the originally-produced pressure. The vendor will explain this by pointing out that some of the values in his calculation are *assumptions*.[31]

Because I have corrupted you into mistrusting all vendor claims, you hire a competent professional to take some readings to quantify some of the assumptions. Subsequently, this competent (and surprisingly good-looking) consultant finds out that your system can only tolerate a 20% reduction in system pressure before it stops functioning properly. Furthermore, he invested all of five minutes to clamp a meter onto the motor and found out that the motor is only 75% loaded, anyway. His analysis looks like this:

Existing			Proposed	
HP	50		HP	50
kW	27.975		kW	20
Hours	8760		Hours	8760
kWh	245061		kWh	178649
$/kWh	0.17			
Existing $	$	41,660.37		
Proposed $	$	30,370.41		
Savings	$	11,289.96		

---

[30] *Totally taking joy in it.*

[31] *Assumptions are dangerous. In the hands of skilled professional, they are necessary evils that fill in the otherwise unquantifiable gaps in any analysis. In certain other hands, they are wild-ass guesses specifically fabricated to make the project look better.*

Well...That's depressing. By assuming the motor was fully loaded, and assuming far too low a minimum speed, the vendor overstated the savings by 410%. In other words, there were 81,687 kWh in the vendor's existing case that never existed in the real world, and 123,899 too few kWh in the vendor's proposed case to do the job.

This Really Happened...

Occasionally, a vendor will so completely misunderstand the nature of VFD savings that truly bizarre assertions will emerge. In one such case, while I was reviewing ECMs for a large Northeast Utility a vendor presented for review a VFD project with many therms in natural gas savings. Here's how it happened.

The vendor, having established that variation existed and that a VFD could be employed, calculated that the variation in heating hot water constituted a reduction in boiler load. It followed the Myth of the Intuitive: less water through the system meant less water the boiler had to heat.

But much like resetting the temperature, varying the flow has no effect on the load in the space. Conversely, you only get to vary the flow *because* there is less load on the space.

It's all part of Thermodynamic Continuity: That flow variation only exists because a valve closed. That valve closed because the thermostat said the space was warm enough. If the space was warm enough, then that space no longer had any load.

If the system had been equipped with a 3-way valve, that hot water would have bypassed the coil and been returned to the boiler with all of its BTUs intact (in the form of warmer return water). Subsequently, the boiler would not have to add those BTUs to the loop. Why would it? They never left! The VFD allows you to send less water, which saves pumping energy, but the result is you do not get that warm, BTU-laden return water you were getting before. But that's OK; because you never had to put those BTU's into the system in the first place.

It's a wash: constant flow loops run more water but return unused BTUs to the boiler. Variable flow loops only add as much water as needed to deliver the BTUs that the load requires. The load on the space stays the same, and the BTUs you have to buy in the form of gas or oil stay the same. The only difference is how much energy you spend pumping them around.

The vendor in question was outraged by my allegation that there were no gas savings and he escalated the issue to my supervisor. When I explained the situation, even a manager(!) could see that there were no thermal savings to be had. No incentive money was given, and that vendor still spits on the ground whenever my name is mentioned to this day.

The key take-away from this example is that VFDs can only save pumping or fan energy, they cannot alter the loads on your space.

When do They Work?

On any system with a motor that has variation or can have variation introduced. Consider the following system:

Let's assume a Boston, MA building has

- Seven large air handlers,
- Each one with a hot water coil and a three-way valve.
- A 50 HP hot water loop pump

We know that the three-way valve's only job is to allow you to alter the flow through the coil without changing the flow through the whole system. If the flow through the whole system doesn't change, then there is no variation.

As far as your vendor is concerned, the scope would have to include replacing those 3-way valves with 2-way valves, in order for the loop to experience variation in flow. This will add cost and complexity, which is why many vendors leave it out. But without it, there are no savings, which makes it an important step!

If we calculated that the minimum speed this pump can run to meet all requirements is 67% (or 40Hz), then over a typical heating season, the savings might look like this:

Hrs	% Heating Load	% speed w/VFD	Pump kW Baseline	kWh Baseline	Pump kW w/VFD	kWh w/VFD
737	2%	67.00%	24.6	18125	7.4	5453
703	11%	67.00%	24.6	17289	7.4	5201
861	21%	67.00%	24.6	21175	7.4	6370
867	30%	67.00%	24.6	21322	7.4	6415
580	39%	67.00%	24.6	14264	7.4	4291
315	49%	67.00%	24.6	7747	7.4	2331
269	58%	67.00%	24.6	6616	7.4	1990
131	67%	67.40%	24.6	3222	7.5	987
56	77%	76.70%	24.6	1377	11.1	622
23	86%	86.00%	24.6	566	15.6	360
5	95%	95.30%	24.6	123	21.3	106
6	100%	100%	24.6	148	24.6	148
0	100%	100%	24.6	0	24.6	0
0	100%	100%	24.6	0	24.6	0
0	100%	100%	24.6	0	24.6	0
		kWh before		kWh after		Saved
		111974	kWh	34274	kWh	77700

Note how the speed never goes below 67%, because then the system would not work!

Now, 77,700kWh is a lot of savings. More than enough to replace seven valves, typically.

VFDs are a powerful and extremely effective tool for reducing a building's energy consumption. There is a reason that they are very popular: They work, they are not huge dollars, and they are not hard to install. Just take care to evaluate your system carefully before writing that check, because the savings can be much smaller than presented.

Myth #7: Outside Air Reset

One of the most common Energy Conservation Measures in our industry is the ubiquitous "outside air reset." Often included in packaged Building Automation Systems (BAS), Energy Management Systems (EMS), or as a stand-alone controls product, these systems purport to save energy by minimizing the temperature of heating hot water and subsequently reducing heating fuel consumption.

While there are several ways in which this strategy does save energy, there is a lot of confusion about how it does so. This leads to much misappropriation of these savings and their magnitudes. Furthermore, the ways in which these controls interact with other building systems is almost never clarified in vendor proposals that include this ECM. This measure often espouses all three types of Energy Efficiency myth.

Outdoor Air Reset (OAR) Defined

As defined in a white paper published by a prominent vendor:

> "Outdoor reset is [a] boiler water temperature control scheme that uses outdoor air temperature to establish the required temperature of the boiler loop needed to keep the building at the desired temperature."

Or another:

> A hot water reset application saves energy by raising or lowering the supply water temperature based on a proportionate drop or rise in outdoor air temperature.

Some vendors will go so far as to claim that OAR will increase the efficiency of the boiler:

Quoted from another prominent purveyor of controls:

Key benefits:
- *Easy installation and wiring*
- *Operate the boiler at the lowest possible temperature to improve efficiency*
- *Minimize ticking expansion noises in baseboards by preventing large temperature swings*
- *Improve the comfort of occupants by better matching the heat requirements based on outdoor temperature*
- *Expand boiler life cycle with the Automatic Boiler Differential feature to minimize boiler short cycling*

Essentially, the boiler controls establish a boiler water or heating loop water setpoint as a function of outside air temperature and adjust accordingly. On warmer days, the boiler water is not as hot as on colder days.

Why does this work?

This adjustment purportedly takes advantage of the relationship between outside air temperature and building loads. This relationship is typically defined by two (hopefully familiar!) basic formulas:

Conductive heat loss: $U \times A \times \Delta T$
Infiltrative heat loss: $1.08 \times CFM \times \Delta T$
Where:
U= U-value of exterior construction
A= Area of exterior surfaces
CFM = Cubic feet per minute of outside air
ΔT = Difference in temperature between indoor temperature and outdoor temperature.

The important part, and crucial to OAR, is the concept of the "ΔT." All heat transfer will proceed from areas of high heat (source) to low heat (sink); and the rate that this occurs is proportional to the difference between the heat source and the heat sink.

On a warm winter day, the outside air temperature is closer to the inside air temperature. This means the ΔT is smaller, and therefore the rate of heat loss to the outside is smaller. When it is colder out, the ΔT is larger (provided the interior temperature is static) and the rate of heat loss is larger. This leads to the very logical conclusion that heating loads on warmer days can be met with lower boiler water temperatures than on colder days.

And they can. That is not the issue, typically.

Problem #1: Load is Load: or the "Myth of the Intuitive"

Remember "Thermodynamic Continuity?" It's very critical here.

Having established that the boiler water temperature can be lowered, many vendors and customers become very excited. It is a very logical and intuitive conclusion that lower boiler water temperatures means putting less energy into the boiler water. This is correct in the sense that it takes fewer BTU's to bring the return water up to 145°F than it does 180°F. If that was all there was to it, then the savings would be quite significant.

Let us consider a space under load and see if it bears out, then.

Assume a building that experiences a heat loss of 10,000 Btu/hr. at OA temp 40°F and interior setpoint of 70°F. In a nutshell, the boiler plant must deliver 10,000 Btu/hr. to hold the setpoint. This same space may experience a heat loss of 20,000Btu/hr. at 20°F, and 30,000 Btu/hr. at 10°F,

but at 40⁰F OA temperature, the load is 10,000Btu *per hour.*

The boiler must supply all 10,000 Btu, within one hour, to hold that setpoint. Now, because it is a balmy 40 outside, let's assume that the controls decide that 148⁰ water is sufficient. It sets the heating loop water to 148 and holds it there. Let's also assume that the controller functions perfectly, the space temperature setpoint is held, and everyone is happy. So how many BTU's did the customer have to buy in the form of gas, oil, propane, etc.?

For every hour at that load, the customer still purchased 10,000 BTU of heating fuel.

Consider the same space without OAR. The boiler sends 180-degree water no matter what the temperature outside is. How many BTUs of 180-degree water did the customer have to buy? The load has not changed, nor have the requirements for holding setpoint, so without OAR, the customer **still had to buy 10,000 Btu per hour.** OAR has not changed the load on the boiler, it has merely changed the rate at which it satisfies that load.

Space load has to be met irrespective of the temperature of the medium employed to deliver the energy. A space that needs 10,000 Btu/hr. will consume the same quantity of heating fuel no matter what temperature the transport medium is.

So what's happening? Enter the fourth dimension!

Still confused? How can reducing the heat content of boiler water *not* save energy? Because there is the fourth dimension to consider: Time. The problem lies not with the Btu's, but with the Btu's *"per hour."*

Here's the catch. The load is immutable. Therefore, if you reduce the energy content per unit of heating medium, then you must use more units of said medium. When you reduce the energy content (temperature) of the water, you have to use more water. How does this translate to our hypothetical space? Back to thermodynamic continuity!

Somewhere in the space will be a heater. This heater will have a hot water coil. This coil will have a valve or a circulating motor that determines how much water the coil receives. Heat transfer from the coil to the space uses the same fundamentals as building heat loss: The greater the ΔT between the coil and the space temperature, the more heat energy gets transferred *per unit of time*. Ergo, when the water is cooler, the coil must operate for more *time* to meet the same load. Either the pump must run longer or the valve must stay open longer to deliver enough energy to the space.

For instance, if our coil is X gpm, and it deposits 500 Btu per minute at 180^0F, then its control valve must stay open for 20 minutes every hour to deliver our 10,000BTU.

If we sent 148^0F water, then the ΔT between the space and the coil will be 32^0 *less* than it was at 180^0. The *smaller ΔT* means *slower heat transfer* from the coil to the space. Now our coil is only delivering 250Btu/min. to the space and must stay open for 40 minutes every hour to meet the 10,000 Btu/hr. load.

Now imagine your hot water loop has a variable Frequency Drive (VFD) on the pump. The OAR controls have created a scenario where your pump must run for longer hours, at higher speeds than if warmer water was sent. This will cause an increase in electrical consumption at a cost that is typically orders of magnitude greater than any savings OAR can provide.

Problem #2: Loss Confusion and the "Myth of Magnitude"

So why do people buy these? Do they really save nothing? Obviously, there are some savings associated with OAR. Notably, vendors will specify three types of savings:

1. By reducing the boiler temperature water, you are reducing the rate of heat transfer to the space, but you are also reducing the rate of heat loss to unconditioned spaces. Every bit of uninsulated pipe, every bit of failed boiler insulation, every leaking valve and hot spot is wasting energy. These are generally referred to as "distribution losses." By reducing the boiler water temp, and subsequently the ΔT, these losses can be reduced. This will lead some vendors to claim that the boiler is "more efficient" with OAR. It would be more accurate to claim that the hot water system has become marginally more efficient, as the boiler efficiency remains unchanged.

2. Pre- and Post-Purge cycles are reduced. By running lower temperatures and keeping control valves and circ. pumps running longer, the boiler will generally cycle fewer times per hour. By holding that lower temperature, you create a scenario where the boiler will run longer, with fewer starts and stops. Boiler purge cycles waste fuel and minimizing these is always beneficial.

3. Reduction in "flywheel effect." When very hot water is sent under low load conditions, occasionally the space will over heat by a degree or two due to residual heat left in the coil (flywheel effect). Sending cooler water reduces this effect.

What's the problem, then? Generally, here are two main issues with the lot of 'em:

1. These are typically very small losses.
2. OAR does not really eliminate these losses, only reduce their frequency and magnitude.

Distribution Losses:

Typical building distribution losses range from 1% on a well-maintained system to 5-6% if the system has been neglected. An absolutely awful building might have 8-12% total system losses. While potentially significant, it is important to understand what your system is actually doing. Many losses accounted for in these estimates are not losses at all. Energy cannot be destroyed. If your boiler made that heat, then that heat is in your building somewhere until it leaks outside. This only happens through an exterior wall or roof.

Many vendors consider any BTU not specifically emitted through a terminal unit to be "wasted." This is a gross oversimplification of what is actually happening. Much of the heat "wasted' in a modern building's distribution system is finds its wat to a conditioned space. Consider an uninsulated hot water pipe in a classroom or in a return air plenum. Heat that escapes these pipes still ends up heating the building. Even if the boiler room is 100 degrees, if there is occupied space above it much of that heat will find its way into that space. This is not to say that all of that heat is recouped, just that not ALL of those BTU's are "wasted."

Vendors also tend to over-estimate the rate at which heat is lost in a distribution system. Pipes in narrow chases or close to the ceiling lose heat at a significantly reduced rate. Since this is where most pipes find themselves, many vendors grossly overestimate the losses associated with them. The international Building Performance Simulation

Association, in a 2013 study, found that typical heat loss estimates, for all of the above reasons, were universally overestimated by vendors[32]:

> *"Results show that SCMs (Simplified Calculation Measures) overestimate the distribution heat losses in these systems."*

> *"...It was found, especially in low-temperature CHD (Combined Heating Distribution Systems), the SCMs largely overestimate the heat losses."*

> *"The few available SCMs for combined space heating [] systems merely reflect a basic combination of the existing methods, rather than an adaptation of the calculation parameters..."*

From "Modelling of Residential Buildings and Heating Systems,"

> *"...measurements reveal significant overestimations of the calculated heat demand and of the calculated global heat losses (transmission and ventilation)."*

Vendors rarely account for all the ways that distribution "losses" find their way back into the space and end up not being wasted at all. When the vendor says your system has "10% losses" it would behoove a building owner to ask for better quantification.

---

[32] *"Heat Loss in Collective Heat Distribution Systems." Eline Himpe, Jluio Efrain Vaillant Rebollar, Arnold Janssens; International Building Performance Simulation Association, August 2013.*

Even if a system is horribly deficient, these deficiencies are far more effectively and inexpensively handled by addressing them specifically, rather than through the installation of boiler controls. OAR does not eliminate these losses, either. It simply reduces their magnitude for a portion of the year.

Pre- and Post- Purge Cycles

Pre and post purge cycles, while wasteful, are typically less than 1% of total heating fuel costs. Furthermore, OAR will never eliminate them, only reduce their frequency under low-load conditions. If your boiler has a chronic short-cycling problem, OAR may be part of the solution, but this is a problem best addressed with reassessing loop properties. OAR will not fix fundamental issues with your heating loop. OAR can actually increase short cycling in some systems.

Flywheel Effect

The flywheel effect typically wastes little to no energy, as every extra degree above setpoint the system experiences translates into a longer period of no heat being added. Essentially, the "extra" BTUs added to the space are not wasted as they still went into the space. They just did so a little too quickly for a minute or two.[33] If this results in occupant discomfort then the flywheel effect can typically be handled by installing a modern, smarter thermostat. Simple controls tweaks can anticipate this issue and modulate valves and circulating pumps intelligently enough to prevent these brief moments of overheating. This is significantly less expensive than OAR, with none of the disadvantages. Algorithms to correct this issue can be

---

[33] *Technically, for this interval, the ΔT is also higher, increasing the rate of heat loss, as well.*

built into an EMS or BMS as well, making this problem a non-issue for most modern installations.

Problem #3: System Interactions and the Myth of the Quantifiable

What else can go wrong? Interactions. Less of an issue when incorporated into an over-arching controls system (BMS, BAS, or EMS), but stand-alone OAR controls can wreak havoc with other building systems.

1. If your hot water plant has a bypass/blending valve and the OAR controls do not account for this, care must be taken with where the sensors are located. If placed incorrectly several negative outcomes are possible. The system may never bypass, or worse, heating loop temperatures may get driven far too low to meet space loads. Furthermore, the boiler water itself might not reset, but rather the loop will (via the blending valve) eliminating boiler-loss savings under these conditions.

2. VFDs will run longer and faster. This is non-negotiable as outlined above. Since VFD savings are typically far larger than any OAR savings, the loss of these will typically result in OAR controls costing the customer money and not saving anything.

3. Hot and cold spots will be magnified. Since the heating water is (typically) a global phenomenon, resetting it to a lower temperature will magnify any deficiencies in the system as a function of load variation throughout the space. Space loads very with fenestration, orientation, condition and configuration. A room that is facing the rising sun will have very different loading than one that is under shadow on the north side of the building.

Rooms with very high outside air leakage may never get to temperature. In older buildings, minimum loop water temperatures may need to be adjusted to a far higher number than initially thought, reducing and even eliminating the benefits of such a system.

4. If your hot water system is not equipped with a bypass loop or a primary/secondary system, holding low loop temperatures can be very bad for the refractory (and other components) of your boiler. Very low return water temperatures can thermally stress the boiler and its components and may even cause severe damage. If your boiler is equipped with the above systems, the boiler may short cycle excessively in an attempt to hold refractory water temperature at a safe level while keeping the loop water cool. Careful commissioning and monitoring may be required.

5. Potable water systems: If your potable water or domestic hot water (DHW) utilizes boiler energy to maintain temperature, resetting the boiler temperature will likely adversely affect your ability to make and hold DHW temperatures. This will necessitate a much higher minimum setpoint, reducing and often eliminating any tangible benefits of OAR.

This Actually Happened...

A customer of a large ESCO was unhappy with the project they had purchased. This project included controls upgrades for about a dozen school buildings in a heating-driven climate. As autumn progressed, the building started to have problems holding temperature. The teachers were mad, the parents were mad, and most of all, the guy in charge of making the payments on the project was mad.

I was dispatched to help trouble-shoot the issue, and I immediately zeroed on the proprietary OAR controls that were installed on the boiler. It was fairly obvious to everyone (including the janitor) that these controls were resetting the loop too cold to meet load. You'd actually be amazed at how many school district custodians make damn fine "practical engineers."

Surprising no one at all, the vendor was adamant that their high-tech devices could not be the problem. They calibrated and re-commissioned extensively, and the problem refused to go away.

As is my way, I went a little bit old school and busted out my graph paper. It always helps me understand a complex bit of plumbing when I sketch it out. If you have never tried to decipher the piping in a 75-year-old middle-school boiler room, I can only exhort you to do it one time in your life. The mental exercise is good for you, and your ability to manage frustration without expletives will be well taxed.[34]

After several hours, I stumbled (entirely coincidentally) upon the issue. The OAR controls installed were "smart" controls that used OA temp and return water temp to determine the correct supply water temperature. Very sophisticated stuff.

Of course, these sophisticated devices were installed on a boiler purchased during the Kennedy administration. The hot water system in question employed a primary-with-bypass configuration. This just means that there is a bypass leg between the supply and return that prevents all of the return water from going through the boiler under low load. A three-way valve blends supply and return flows to meet

---

[34] *And the smell of asbestos is quite bracing.*

the load without running the boiler too much or thermally shocking it.

The new OAR controls were resetting the loop temperature at the boiler, and the bypass valve was resetting the temperature from the return side. The result was that as one system was resetting the temperature, the other system tried to correct it. When the OAR control reset the supply water temperature, the three-way valve would then route more water through the boiler (because it wasn't hot enough). Returning more warm-ish water to the boiler caused the control to perceive the loop as lightly loaded, which made it set the temperature even lower. This feedback loop caused the main loop temperature to keep getting colder and colder until there was not enough heat in the loop to hold temperature.

Naturally, the OAR controls had to be removed and the customer given a full credit for that part of the job...

Yeah, right. The OAR controls were simply set to a minimum temperature that was so high, it could not possibly interact with anything else. The device saved precisely 0 dollars for the customer and I was once again accused of "not being a team player."

So When is OAR a Good Idea?

None of this is to say that OAR is not a "good" ECM. It has several key advantages and when properly applied can be very beneficial. OAR is most effective when you have a constant volume, high distribution-loss system with consistent loads and minimal space control. If your system has very little in the way of individual control valves or thermostats, OAR can often be an excellent way to control losses. Small domestic-style systems, with few zones and utilizing hydronic baseboards, are typically the best systems upon which to employ this measure.

Simple primary systems, if they can safely handle lower return water temperatures, may enjoy some savings with OAR. Operators must take great care to make sure that lower return water temps will not damage the boiler. Tread carefully.

Under these or similar conditions, a well calibrated boiler control system incorporating OAR can be a great tool to control heat loss costs. It is a proven and effective measure when properly applied and commissioned.

This measure falls under all three myth types. While large savings feel like they should be there, they just typically aren't. Worse, you can seriously break your building if you don't pay close attention to what is going on.

Myth #8: Air-Source Heat Pumps and Why We Hate Geography

Heat pumps are amazing machines, and their ability to operate at high efficiencies has made the market very receptive to their implementation. Early adoption of this technology has been a benefit for both vendors and consumers.

Most of the time.

What is a Heat Pump?

Heat pumps come in many configurations. Air-source heat pumps, water source heat pumps, even ground-source heat pumps ("geothermal") are all enjoying increasing popularity.

All heat pumps follow a sequence called the "Carnot Cycle." What's the Carnot Cycle? Let me just search for that... Here you go:

Highly Nerdy Stuff Here:

The Carnot Cycle when acting as a heat engine consists of the following steps[35]:

Reversible isothermal expansion of the gas at the "hot" temperature, T1 (isothermal heat addition or absorption). *During this step the gas is allowed to expand and it does work on the surroundings. The temperature of the gas does not change during the process, and thus the expansion is isothermal. The gas expansion is propelled by absorption of heat energy Q1 and of entropy $\Delta S = Q\_1/T\_1$ from the high temperature reservoir.*

---

[35] *Wikipedia*

Isentropic (reversible adiabatic) expansion of the gas (isentropic work output). *For this step the mechanisms of the engine are assumed to be thermally insulated, thus they neither gain nor lose heat. The gas continues to expand, doing work on the surroundings, and losing an equivalent amount of internal energy. The gas expansion causes it to cool to the "cold" temperature, T2. The entropy remains unchanged.*

Reversible isothermal compression of the gas at the "cold" temperature, T2. (Isothermal heat rejection)) *Now the surroundings do work on the gas, causing an amount of heat energy Q2 and of entropy \Delta S=Q_2/T_2 to flow out of the gas to the low temperature reservoir. (This is the same amount of entropy absorbed in step 1, as can be seen from the Clausius inequality.)*

Isentropic compression of the gas (isentropic work input). *Once again the mechanisms of the engine are assumed to be thermally insulated. During this step, the surroundings do work on the gas, increasing its internal energy and compressing it, causing the temperature to rise to T1. The entropy remains unchanged. At this point the gas is in the same state as at the start of step 1.*

Well. That was horrible now, wasn't it?

How about this:

> *"Heat pumps use expanding and contracting gasses the same way your refrigerator does to move heat around."*[36]

That's a whole lot easier, isn't it?

Fundamentally, a heat pump is a device that transfers heat from one area to another. Technically, your air conditioner is a heat pump: it takes heat from inside your building and

---

[36] *Andrew "Brevity is the Soul of Wit" Vaillencourt*

moves it outside. If you turn it around it will pump heat from outside into the inside just as easily. It's why the coils on your refrigerator are hot. They are taking the heat out of the inside of the box and putting it outside.

A critical piece of information for this type of system is that they do not "create" heat the way combustion does. They take existing heat, heat that is already there, and move it to somewhere else. This requires much less energy than combustion, which rips apart chemical bonds in an exothermic reaction. In the case of Air-Source heat pumps, that "somewhere else" is "outside." This presents a unique set of challenges.

Problem #1: The Ratings Game and the Myth of Magnitude

Heat pumps are typically rated in two ways: EER and COP[37]. If you took the time to read the definitions in the beginning of this book, then you remember that:

EER- (Energy Efficiency Ratio)

The ratio of output cooling (in Btu/hr.) to input electrical power (in watts) at a given operating point.

COP- (Coefficient of Performance)

The ratio of heating or cooling provided to electrical energy consumed.

Both are measuring the same thing: How much energy I have to put in to get something I want out. For instance, if I have to put one kW into my heat pump to get 3.2 kW worth of heat out of it, then the COP is:

3.2(kW output)/1 (kW input) = 3.2
COP=3.2

---

[37] *As well as SEER, HPSF, kW/ton, and several others.*

For the same system, the EER would be the same ratio in BTU, or:

$$(3.2kW*3.412\ Btu/W)/1kW = 10.92$$

$$EER=10.92$$

So, where's the Myth? Both of these ratings are ratios, which means they follow a nice linear correlation. If the universe worked that way, this section would be superfluous. But the universe is fickle, and the reality is that these ratings are for one discreet set of conditions only. For instance, the EER is typically measured at:

	Dry bulb temperature	Wet bulb temperature	Relative humidity	Dew Point
Outdoor conditions	95°F (35°C)	75°F (24°C)	40%	67°F (19°C)
Indoor conditions	80°F (27°C)	67°F (19°C)	51%	60°F (16°C)

The uncomfortable truth is that the values of these ratings will change as conditions change. This sort of makes them useless as ratings, but no one really likes to mention that. Every time I bring it up people scowl and call me "difficult to work with.[38]"

Many vendors will advertise an air-source heat pump with a heating COP of 3. When you consider that the COP of a typical boiler is .85, this is very compelling. But remember how a heat pump "makes" heat? That's right: It doesn't. It moves heat from one place (source) to another (sink...or 'load,' in this case). In the case of an air-source heat pump, that heat source is the outside air. So riddle me this, dear reader: How much heat do you think is available when the outside air temperature is 10°F?

I'll give you a hint: *Very little.*

---

[38] *I'm starting to think I'm not a very likable person.*

How easy is it, do you suppose, to extract heat from 10ºF air?

Another hint: *Not easy at all.*

And finally, how much heat do you typically need when it is 10ºF outside?

Last hint: *A lot.*

Let's recap. When it is very cold outside you have a heat source that ain't all that hot. Which makes it difficult to extract any heat from it. Simultaneously, this is occurring during the hours when you need heat the most. Does any of that sound like a recipe for awesomeness?

Here is a visual aid that graphs the predicted kW input required to meet a heating load as a function of Delta T (indoor setpoint minus outdoor temperature) using the rated COP and the actual tested and metered energy input.

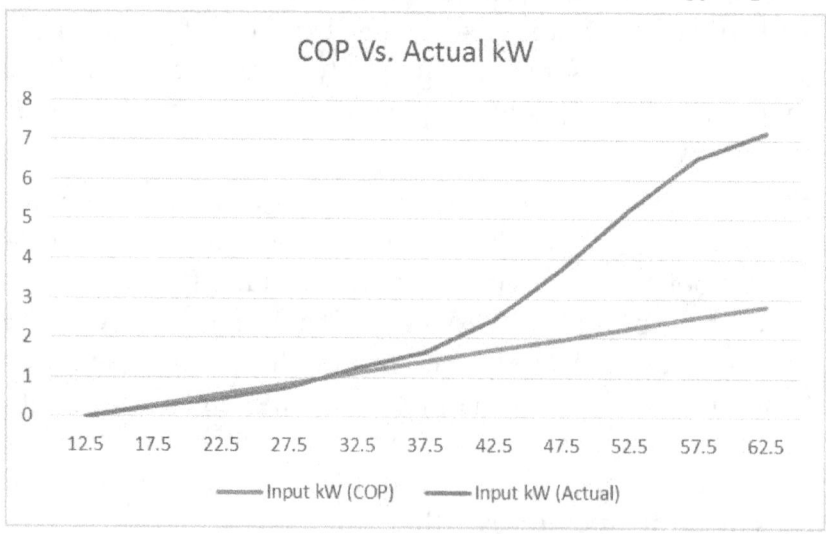

As the Delta T increases (x axis), the predicted input kW climbs (y axis), as we would expect it to. Increased Delta T is, by the very definition of the concept, increased load. Increased load means you have to put more heat into the space to meet it.

But look at those two lines start to diverge around 37.5 degrees delta! If you were to use rated COP to calculate your heating kW, you would ultimately get a very wrong answer over any reasonable spread of outside air temperatures and Delta T's.

Why is this done this way? It's an easy, simple system, that's why.

Most COP ratings are for outside air temperatures at or above freezing. Therefore, a COP of "3.0" on the box means a COP of "much less than that" at very cold temperatures (cold temperatures mean high Delta T!). If you live in a fairly warm climate, then the damage is minimal. Below the Mason-Dixon Line, air-source heat pumps are often a very good way to save energy.

If you are in New England, or Montana, or Alaska? Well, here is what is going to happen. When it gets cold outside, your heat pump is going to try very hard to extract heat from that ice-cold air. When you extract heat from something, it gets colder. When that something is already cold there is less heat to extract. Now the compressor has to use an increasing amount of energy to extract a diminishing quantity of heat. The compressor runs longer and harder while pumping less heat, bringing the COP down. As the temperature drops, our heat pump will find itself using more and more electricity to do less and less of the work you need it to.

Oh yeah, it will also freeze that sucker solid as a rock. Of course, manufacturers aren't stupid, and your heat pump probably has an electric heater installed to thaw the coils and supplement the heating capacity. Electric resistance heat has a COP of 1. That means it uses more energy per unit of heat than the compressor, making the COP go down even further. Some newer heat pumps will cycle between rejecting and absorbing heat to keep the coil from freezing, and they will proudly advertise that they are still heat pumps down to 0°F. But that constant back and forth uses more juice too; typically to an extent that the COP at 0°F is so close to 1 that the distinction is irrelevant.

"Wait a minute," I hear you thinking, "These things will cycle between heating and cooling? Does that mean that when it is really cold outside, they will actually use indoor heat to unfreeze themselves?"

It sure does. What a great observation!

"But doesn't that mean that the actual *heating capacity* of the heat pump shrinks as it gets colder?"

Yup, the spiral is inescapable: the colder it gets outside, the less heat the darn thing pumps, for the same amount (or more) of juice. Here is a graph of a prototypical heat pump performance curve:

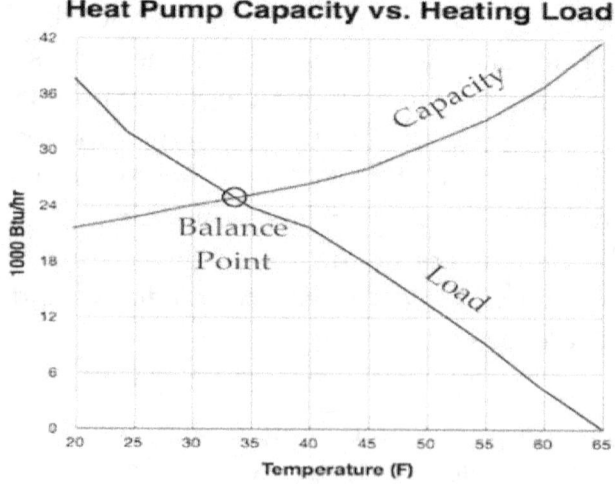

For any outdoor temperature below that balance point, you will need supplemental heating, or your space will get cold. Where that point occurs will vary with the capacity of the unit and the nature of the space, but they all have a balance point that you had better have a handle on before you write that check.

If you want to actually keep your space warm when it is cold outside, you have to buy a much larger unit than most people realize. If you are thinking that bigger units mean bigger dollars, then congratulations! You are starting to get the hang of this.

Now let's have some fun. On the next page are the actual, advertised and certified ratings for a commercial 10-ton (120,000 BTU) air-source heat pump. The manufacturer and model are irrelevant, just understand that this is actual product data from a name-brand vendor, and it is considered to be a very high-performing (expensive) model. This particular unit does not employ an electric heater, but rather cycles the unit back and forth between heating and cooling to prevent freezing. This is done intelligently and carefully to maximize low-temp COPs and keep the unit as efficient as possible at all temps. It is a legitimately nice model, and one I have recommended to customers in the past.

Space setpoint (deg F)

OA Temperature	MBTU capacity	60	70	75	80
-18		29.9	42.3	46.5	49.9
-13		39.4	47.5	50.6	53
-8		46.9	52.6	54.8	56.7
-3		53.7	57.8	59.4	60.7
2		60.3	63.1	64.2	65
7		66.9	68.6	69.2	69.7
12		73.4	74.3	74.6	74.7
17		80.2	80.3	80.2	80.2
22		87.1	86.6	86.3	86
27		94.2	93.2	92.6	92.2
32		101.7	100.2	99.5	98.9
37		109.1	107.2	106.4	105.6
42		116.7	114.4	113.5	112.7
47		124.6	122.1	121.1	120.1
52		132.8	130.1	129	127.9
57		141.1	138.3	137.1	136
62		149.6	146.7	145.3	144.1
67		158.3	155.2	153.8	152.5
72		167.4	164.1	162.6	161.2

This unit may be rated for 120,000 BTU at 45 degrees, but when it gets down around zero? You only get 63,000 BTUs if you want it to be 70°F in the space. That's half the advertised capacity.

At least it's efficient, right? Let's look at the COP at the various temperatures:

OA Temperature	Space setpoint (deg F)	60	70	75	80
-18	**COP**	1.2	1.5	1.5	1.5
-13		1.5	1.6	1.6	1.6
-8		1.7	1.8	1.7	1.7
-3		1.9	1.9	1.9	1.8
2		2.1	2.0	2.0	1.9
7		2.3	2.2	2.1	2.0
12		2.5	2.3	2.2	2.1
17		2.6	2.5	2.3	2.2
22		2.8	2.6	2.5	2.4
27		3.0	2.7	2.6	2.5
32		3.2	2.9	2.8	2.6
37		3.4	3.0	2.9	2.7
42		3.5	3.2	3.0	2.8
47		3.7	3.3	3.1	3.0
52		3.9	3.4	3.3	3.1
57		4.0	3.6	3.4	3.2
62		4.1	3.7	3.5	3.3
67		4.3	3.8	3.6	3.4
72		4.4	3.9	3.7	3.5

You can see how the COP changes with the indoor air setpoint and the outdoor air temperature. At an indoor temp of 70°F and an outdoor temperature of -18°F, this unit has a COP of 1.5. That's markedly better than your boiler on efficiency, but it is less than twice as efficient at several times the fuel cost (see next section) to operate.

This example is for a very nice piece of tech. Chances are, you aren't dealing with something this well-designed. For most air-source heat pumps, once it gets real cold outside the electric heater comes on and stays on. This turns your expensive heat pump in to an expensive electric heater with a COP of 1. This will have a savage effect on any energy savings calculation and is often omitted from vendor proposals.

Below 25°F, most small (under 7.5-ton) heat pumps are really just electric strip heaters. If you live in Boston, then you have about 1,100 annual hours below 25°F. You can imagine what Buffalo New York, Glasgow Montana, or Portland Maine, get.

Problem #2: Money is Green, Too!

But hey, a COP of 1 is still better than a COP of .85 like my boiler has, right? So I am still ahead of the game!

That depends on your definition of "ahead," actually. At $1.10 a therm, a gas boiler with a COP of .85 costs about 13 dollars for every usable million BTUs it produces. With an electric cost of $.16/kWh, an air-source heat pump wheezing along on the electric heater at a COP of 1 costs about 49 dollars for the same million usable BTUs. So even though you are being more "efficient," your heating costs are nearly quadruple what they would be on gas when the outside air is below a certain threshold. Most small commercial heat pumps end up having to run the electric resistance coil full time somewhere around 25 degrees OA temp and below. Better ones will run defrost cycles and hang in there for a few more degrees, but even then, their COPs are brutally low.

So, heat pumps almost always save energy. That's not the problem, here. The myth is that they will save you a lot of money. If you are far enough south, then they can be very good. If your building is somewhere with a fair amount of sub-freezing hours, they can be a very expensive way to save energy. If the vendor calculates the savings at a COP or EER that only reflects favorable conditions, then the heat pump will end up saving plenty of energy while costing you a massive amount of money. Saving energy is a good thing. We all want to be "greener." But, hey, money is green, too!

This Really Happened...

A particular ESCO I once worked for[39] took a job at a meeting hall and club house that needed help with their energy bills. Citing the fantastic heating COP of a high-dollar air-source heat pump, this ESCO removed the old, inefficient hydronic heating from several banquet rooms and replaced them with these heat pumps.

Doing so greatly increased the efficiency of both the heating and cooling in these spaces, and all involved were extremely thrilled with their work. There were handshakes and high-fives a-plenty.

Until the winter set in, and those heat pumps became electric heaters. The client's utility costs nearly doubled, and it was many a grim and angry phone call to the ESCO. There were recriminations, and accusations, and much wailing and gnashing of the teeth.

This ESCO, still not understanding what could be wrong, hired a consultant to evaluate where the savings had escaped to. Very slowly, and carefully choosing his words, the consultant gingerly explained that the old hot water heat, though inefficient, was 70% less expensive to operate than the heat pump.

To this day, I am not sure how the issue got resolved, but I assume there was very little return business from this customer.

---

[39] *None of this occurred while I worked there!*

When is an Air-Source Heat Pump a Good Idea?

There are two primary scenarios were air-source heat pumps really shine:

1. Is your building located between Southern Pennsylvania and the Tropic of Cancer? If so, you are in a good place for an air source heat pump. As a born and bred New Englander, what most of y'all would call a "heating hour" I consider beach weather, and this is key for air-source heat pump success. As discussed, there is a limited outside air temperature window where the COPs are high enough to outperform fossil fuels for heating costs, but cool enough to require heat. Warmer climates have relatively few heating hours, and those hours typically fall within the sweet spot for a good COP.

2. Are you already stuck with electric heat? If your building is on electric resistance heat already, then you are getting destroyed every heating season by those insane electricity costs. Going to a heat pump will at least give you a few hours every year where you can enjoy the benefits of a high COP. Even the chilly north has shoulder seasons with a decent quantity of "sweet spot" hours for an air-source heat pump. You are still going to get blasted in the deep winter, but a little relief is better than none at all.

Myth #9: Condensing Boilers Are Always 98% Efficient

Condensing boilers are excellent technology. Let's get that out of the way now. I love the things. Typical vendor marketing will point out at length that these boilers are (up to) 98% efficient, and that is a whole lot better than what you've got now! At this point in the book you know better than to get excited, and that is good. Soon you will be as jaded and cynical as I am, which means I win.

Problem #1: It's in the Fine Print (Myth of Magnitude)

A boiler takes fuel (gas/oil/biomass or whatever) and ignites it, which is to say it rapidly introduces the process of oxidation. This rapid oxidation (called "burning") of the fuel is highly exothermic and gives off heat.

Boilers heat spaces by burning fuel and transferring the energy of combustion to water. Pumps take that hot water and move it to the space, where it gives away that accumulated heat to those in need. We use water to transfer that heat energy because transferring the actual flame is difficult and dangerous. Surprisingly fun, I concede, but dangerous nonetheless.

The process of oxidation is not perfect. Not all of the fuel gets burned (we will discuss this later), and not all of the heat finds its way into the water. Some of the heat carries the by-products of combustion (CO_2, water vapor) up the stack, and some of the fuel just doesn't oxidize. Nature rarely allows for a perfect stoichiometry in chemical reactions.[40]

---

[40] *Stoichiometry /ˌstɔɪki'ɒmitri/ is the calculation of relative quantities of reactants and products in chemical reactions.*

Typical, no frills, nuthin'-special natural gas boilers can reliably operate at 85% combustion efficiency. Which is to say that 85% of the fuel that goes in to the burner comes out as heated water. The other 15% goes up the stack as flue gas. Consider it a tax on your heating that has to be paid to the gods of chemistry and physics.

Now, that flue gas (or stack gas) has heat in it. That is heat that you bought. That is your heat, and it is going up the chimney. If we could extract that heat, and put it to use, that would be a very good thing. There are a few ways to recover this heat, but in this chapter we are after the most common one: Enter the condensing boiler.

A condensing boiler runs the flue gasses through one end of a heat exchanger and the return water through the other. As the heat leaves the flue gas and enters the return water, the gas, cooling down below its dew point, "condenses," and some moisture and particulates fall out of the gas. This is why it is called a condensing boiler. Just like in refrigerators and heat pumps, condensing a gas gives up a quantity of heat called the "latent heat of vaporization.[41]" This transfer of heat allows the accumulated heat in the flue gas to pre-heat the return water. If the return water is warmer, then the boiler has to put fewer BTUs into it to heat it up to temperature. Under the right conditions, this will create a scenario where 98% of your fuel is converted into usable heat.

But that's the problem, isn't it? "Under the Right Conditions" has been a thorn in the side of engineers since the invention of the wheel. What are those conditions?

Low-load conditions. Here's why:

---

[41] *Latent heat of Vaporization is the enthalpy change required to transform a given quantity of a substance from a liquid into a gas at a given pressure.*

Heat only moves one direction, remember? When the load on the boiler is high (it's really cold out) the space needs a big ΔT between the space and the coil to ensure enough heat gets transferred in a short enough period of time. Under these conditions, the condensing boiler must send 180-degree water to meet the load. If it sends water that's cooler, the load may not be met fast enough, or at all. Ergo, the boiler sends the hottest water it can. That water then gives up a bunch of BTUs to the space and returns to the boiler much cooler. Unfortunately, because it was so hot to begin with, it will not return cool enough to condense that flue gas. If the gas does not condense, then the latent heat of vaporization cannot help preheat the return water.

When the loads are relatively small, the condensing boiler can send cooler supply water, which results in *very* cool return water. It can then then employ this very cool return water to condense away to its heart's content.

Consider a New England building (We will need some good heating hours!) with a 1 million BTU natural gas boiler at 85% efficiency and 85% loading. They want to replace it with a new condensing boiler because the salesman has promised them in increase in efficiency of 13% with his 98% efficient product.

We'll go ahead and use Boston weather data, (available at the NOAA website) and plug in a typical, nothing crazy linear load profile that starts at zero load and tops out at 85%. Which is to say that on the coldest day of the year (a "design day" in HVAC-talk) 85% of the boiler capacity will be required to hold space temperatures. The vendor, in an attempt to inflate the baseline might claim the boiler is 100% loaded...but since you have your actual gas bills you know that is a silly assertion. With threats of grievous bodily harm you force him to use 85% because you know what you are doing.

The savings calculation from the vendor might look like this:

OA	Total	Heating	Existing		Proposed	Eff.	Gas Therm
Temp	Hours	%load	Gas Therm		% load	@ load	@ load
97.5	0	0%	0.0		0%	98%	0
92.5	7	0%	0.0		0%	98%	0
87.5	73	0%	0.0		0%	98%	0
82.5	252	0%	0.0		0%	98%	0
77.5	388	0%	0.0		0%	98%	0
72.5	710	0%	0.0		0%	98%	0
67.5	693	0%	0.0		0%	98%	0
62.5	981	0%	0.0		0%	98%	0
57.5	845	0%	0.0		0%	98%	0
52.5	646	0%	0.0		0%	98%	0
47.5	687	4%	298.7		4%	98%	259
42.5	679	11%	885.7		11%	98%	768
37.5	905	18%	1967.4		18%	98%	1706
32.5	738	26%	2246.1		26%	98%	1948
27.5	491	33%	1921.3		33%	98%	1666
22.5	274	41%	1310.4		41%	98%	1137
17.5	269	48%	1520.4		48%	98%	1319
12.5	77	55%	502.2		55%	98%	436
7.5	33	63%	243.9		63%	98%	212
2.5	9	70%	74.3		70%	98%	64
-2.5	3	78%	27.4		78%	98%	24
-7.5	0	85%	0.0		85%	98%	0
	4165		10,998				9,539

That's 1,458 therms saved and about $1,605 dollars a year. Feels like real money to me! But since we know that it cannot possibly be 98% efficient all year round, we have to consider that. As a building owner, you run your own calculation and apply your knowledge of boilers to adjust the efficiencies based upon the relative system load. Or you hire someone to do that for you. Whatever, it's your building. Your numbers might look like this:

OA Temp	Total Hours	Heating %load	Existing Gas Therm	Proposed load	Eff. @ load	Gas Therm @ load
97.5	0	0%	0.0	0%	98%	0
92.5	7	0%	0.0	0%	98%	0
87.5	73	0%	0.0	0%	98%	0
82.5	252	0%	0.0	0%	98%	0
77.5	388	0%	0.0	0%	98%	0
72.5	710	0%	0.0	0%	98%	0
67.5	693	0%	0.0	0%	98%	0
62.5	981	0%	0.0	0%	98%	0
57.5	845	0%	0.0	0%	98%	0
52.5	646	0%	0.0	0%	98%	0
47.5	687	4%	298.7	4%	98%	260
42.5	679	11%	885.7	11%	97%	779
37.5	905	18%	1967.4	18%	96%	1745
32.5	738	26%	2246.1	26%	95%	2010
27.5	491	33%	1921.3	33%	94%	1737
22.5	274	41%	1310.4	41%	93%	1195
17.5	269	48%	1520.4	48%	92%	1401
12.5	77	55%	502.2	55%	91%	467
7.5	33	63%	243.9	63%	91%	229
2.5	9	70%	74.3	70%	90%	71
-2.5	3	78%	27.4	78%	89%	26
-7.5	0	85%	0.0	85%	88%	0
	4165		10,998			9,919

Note the efficiency of the boiler shrinking as load increases.

That's only 1,079 therms saved and $1,187 dollars a year. It looks like your vendor has overstated the savings by 35%. In engineering terms, an obvious miscalculation of 35% is technically referred to as a "job-killer."[42]

---

[42] *Or at least it ought to be.*

Problem #2: Payback is a…well, you know…

The second problem has nothing to do with efficiency. It has to do with payback. Simple payback, that is. These boilers are expensive. They are high-tech and complicated. Prices will vary, but you can easily expect a condensing boiler installation to be anywhere from 30% to 60% more expensive than a cheap boiler. Getting the simple payback under 15 years is tough to do under these conditions.

Why so pricey? They have to be installed by someone who knows how they work. You really cannot afford to skimp on this. Anyone can bolt a condensing boiler to the floor and run the plumbing. But minor oversights can cause big problems. You want examples? OK:

- Say, for instance your DHW heater runs off your boilers, you have pretty much eliminated all the condensing hours from your year. Your DHW needs a fixed temperature from the loop.
- Your installer may plumb the flue to your existing stack. This is no big deal as long as poisoning your entire building with carbon monoxide and other toxic gasses is okay with you. Since the heat has been extracted from the flue gasses, they do not rise as much. When condensing you can expect your flue or stack gasses to be at about 90 degrees. That's pretty chilly for stack gas. That means the gas may cool off too much and may not rise all the way to the top of the stack. If that happens these gasses will sit there and collect in the stack until they choke the burner completely or leak into the space and kill everyone in it.[43] Most installers know better, but it

---

[43] *You've got CO detectors, right?*

has happened before. Typically, the boiler stops working before the fatalities start anyway.

- If you burn heating oil, you will need special filters and drains because heating oil contains sulfur, so when you condense heating oil exhaust you get hot sulfuric acid. Concentrated sulfuric acid. The good stuff. This will destroy your boiler very quickly. Kits are sold to neutralize and drain it. If you don't know you need one, or worse, you "value-engineer" it out, you are going to have a bad time.

All of this contributes to making the installation difficult and expensive.

This Really Happened...

At one point, I was the senior engineer for a small-ish ESCO and was tasked with assisting sales personnel in quantifying savings potential for the customers they were soliciting. One salesperson was perennially recommending condensing boilers for customers that could not possibly afford them, or on systems that would never realize enough savings to justify their existence.

We came to loggerheads after the fifth or sixth proposal of his that had to be discarded because his savings estimates were 30-50% higher than reality would dictate. This was obviously my fault.

This salesman was convinced that all existing boilers were never more than 70% efficient, and condensing boilers were 95% efficient under any and all conditions. No amount of reason, science, or math could convince him otherwise. He was positively furious with me for "killing the sale" every time he recommended one. Fortunately for me, I do not feel emotions[44] and his fury was mostly amusing to me.

The only real lesson from this story is that this was a real salesman, at a real company, plying his trade on real customers. He was not unique. He is still out there selling. You may just have to deal with him someday. There are lots of these guys out there. Be advised.

When are they a good idea?

If you burn natural gas and you can afford one, these are always a good idea. The savings are not gigantic, but they are real and consistent. There is plenty of engineering magic to play with here.

Otherwise; these all serve well:

- As the supplemental heat source for a water-source heat pump loop.

- As a stand-alone DHW heater (set below 130°F).

- In any process or HVAC application where the return water temperature is below 130°F.

This is a very solid ECM, it just doesn't pay back very fast. That should not deter you from exploring it. Especially if you are in the market for a new boiler anyway. If we exclude the costs that would be associated with any boiler installation, the extra costs associated with a condensing boiler will likely have a payback in the 2-5-year range, and that is a very good investment.

---

[44] *My capacity for human emotion was removed to make room for more math. Right around age 12.*

Myth #10: Energy Management Systems (EMS) Save 10%

Let's start this one correctly: Energy Management Systems are the single most powerful tool a building operator can have to control energy consumption and costs. Depending on how sophisticated a system we are talking about, no single ECM has more potential to make your life as a building operator easier and more efficient than a good EMS.

So why are they here? Because once again the industry is getting a little out of hand with the marketing of these systems.

Essentially, an EMS[45] is any system that monitors and controls your building equipment. Modern versions are vast intricate networks of sensors and controllers that all feed back to a central panel and bring the data to a computer (or website, or IP address). This is called the "head end" because it is where the operator can monitor the building and make changes and adjustments. This type of EMS is called Direct Digital Control (DDC). They are very nice.

Older versions are typically equally vast networks of pneumatic tubing that criss-cross the building sending different pressure signals to different devices to alter their behavior. The head end for a pneumatic system is usually a huge panel in the boiler room with lots of little knobs, dials, and valves that an operator adjusts manually to make system changes. This is called a Pneumatic system (clever, I know) and they are less nice.

---

[45] *Or BAS, or BMS or whatever acronym your vendor is using.*

Then there are the hybrids. These are old pneumatic systems that have had electronic controls "overlaid" in a manner that allows a computerized head end to manipulate the knobs and valves in such a way as to allow for the central computer to exert control over the building. They tend to be enigmatic, quixotic, and misshapen things. They typically have intricacies and idiosyncrasies known only to the installing technician and the most knowledgeable of building operators. But when handled by pros, they usually work just fine.

Nowadays we also have completely wireless, thermostat-based systems that allow users to make changes and get alerts over the internet or internal networks. These are really just fancier versions of a programmable thermostat and only give the user limited control of thermostat signals. This is not the same as a "real" EMS that actively and directly controls the equipment. But these can be comparatively inexpensive and give operators access to the most important and common HVAC functions. If you are broke, these can do a perfectly acceptable job.

Problem #1: Rules of Thumb Only Work for People who are all Thumbs.

The most common myth associated with this is a classic "Myth of the Quantifiable." Since an EMS controls lots of different things, it can DO lots of different things. That sounds great, but building systems are holistic. Everything you do effects everything else.

For instance:

- If you change the setpoint in room A, it will alter the loads on Room B if they share a wall.
- If you change the discharge air temperature on an air handler you have altered the loads for every

piece of equipment that supplements that air handler.

- If you reduce fan speed, then you have changed the flow characteristics for the whole system.

None of this is necessarily bad, but it makes predicting the savings associated with a new EMS very difficult. The time and effort it takes to produce a defensible calculation model for an EMS is significant and requires a pretty good understanding of the systems in question.

Take a good hard look at your vendor. Look at the proposal. Does it look like a lot of hard work from people knowledgeable in your building's systems spent hours developing a robust prediction engine? Were all the individual effects from all the pieces of equipment parsed out and calculated? Are there several pages of calculations in the appendix of the proposal?

Unless you are very large customer, the answer is "probably not." Furthermore, we can't get too mad at the vendor for not wanting to do it. Doing all of this stuff takes time and money. It's an expensive pain in the neck for the vendor, and if they haven't even made a sale yet, why would they bother?

But they have to tell you something, right? So they go back to the "rule of thumb." The industry rule of thumb for EMS installations is that they will save the average facility 10% of their utility bills. As rules of thumb go, it's not terrible. But it is no substitute for real research and analysis. Many EMS's save far less than that, and plenty of them save a lot more.

EMS's are expensive items; and not knowing the potential for savings means rolling the dice with a lot of capital. You would be well advised to analyze the potential very closely before agreeing to purchase one. Reputable vendors will do the math and show their work.

If your vendor doesn't want to show their work, or do the hard math, it might behoove you to look for another vendor.

Problem #2: "Newer" Does Not Always Equal "Better."

Now we venture into the "Myth of Magnitude."

The main energy-saving benefit of an EMS is the ability to schedule when you do and do not put energy into the space.

Primarily, this takes the form of "setback," where space temperature Setpoints are "set back" during the hours when they are unoccupied. For instance, after 6PM an office building may decide to reduce the wintertime indoor setpoint to 55^0F. This way, the ΔT between indoor and outdoor temperature is much smaller, and heating loads are significantly reduced. Conversely, in the summer, the setpoint might be "set back" to 90^0F to reduce the cooling load.

Everyone who has ever turned their thermostat down at night gets this. It's a darn good idea all around, and it has been in use for decades.

Therein lies the crux of the issue. Many commercial buildings built throughout the late 70s and 80s will likely have some sort of EMS already. Not all of them, not even close. But a significant enough proportion that the problem that starts to crop up is this:

If a previous EMS remains, and is functioning as designed, the building is almost certainly already setting back. Installing a new, expensive, and "better" EMS may not generate any more savings of that is the case. If your vendor used the rule of thumb and assumed a 10% savings, then you are going to be very disappointed when you get your utility bills after the fact.

This issue is most prevalent when a vendor comes in and wants to replace your old pneumatic system or hybrid system with a new, all-digital EMS. There are many advantages to a Direct Digital Control (DDC) EMS, of course. Here are some:

- They are far easier to use.
    o They are typically graphically based and have very intuitive interfaces.
    o The results of changes made are immediately obvious.
    o No special HVAC skills are typically necessary to run the schedules.
- A much larger variety of communication options is available.
    o You can receive e-mail, text, or phone alerts for issues and problems, and make changes from anywhere with an internet connection.
    o They will typically have many more sensors and data points, giving the operator a much better idea of how the building is performing.

All of this is very good stuff. However, if your old system is taking the building in and out of occupied mode on schedule already and is already making such changes as are advantageous, then none of these features inherently improve the energy savings associated with that. The savings for the new EMS are contained entirely in whatever new ECMs (if any) are incorporated into the new controls. Savings for those ECMs must be enumerated and calculated individually, which is hard.

Problem #3: The "M" is for "Management."

The most common reason an EMS fails to deliver the expected savings is not the vendor at all. It's the operator. Thanks to "rule of thumb" thinking, it is not uncommon for the assumption to be that the mere presence of an EMS is what generates the savings.

This is NOT the case. *An EMS does not create savings; it allows the operator to create savings.* It gives the building operator the power to make changes to how much energy the space receives by actually changing the system parameters. It allows the operator (amongst other things) to change space temperatures, schedule equipment, alter fan speeds and loop temperatures. It puts the operator in charge of the building equipment. The EMS forces the building to behave in the way that the operator tells it to. This creates a problem that is hard to convey subtly, so I shall do so without subtlety:

An EMS has to be operated by a competent individual.

All that power and control is magnificent for matching your energy consumption to the needs of the facility, but it also means that haphazard manipulation by staff that do not respect the holistic nature of building systems will result in a whole lot of trouble. If an EMS gives the operator the ability to do things that save energy, then the operator must DO THOSE THINGS.

An EMS is a tool, and a tool is only as effective as the wielder. The vendor is likely to extol ad nauseam the ease and convenience of their product. They will point out the intuitive and graphically impressive control panel screen. They will do a very smooth and rehearsed demonstration of the product and they will assure you that anyone can operate their system. For the most part, they will be right.

Fine, let's explore that, shall we? Anyone can operate an EMS? Imagine yourself at Indianapolis Motor Speedway, staring at a 1.5 million-dollar Formula One race car, and you have been informed that you will be competing in tomorrow's race. It's kind of daunting, right? But realistically, it's just a car and you know how to drive a car, right? The pedals are where you expect them to be, the steering wheel is still a steering wheel, and the various buttons are clearly marked. How hard can this be? With a cavalier shrug you hop in and head for the races under the assumption you are ready to run with the big boys.

But let's be honest with ourselves, here. The best possible ending for this story is that you manage to get around the track without dying and accomplish the impressive task of coming in dead last. Worst case scenario? Lots of flames and a horrific yet spectacular demise. It does not take a large deductive leap to understand that driving your '09 Camry is not the same thing as racing a Formula One car, does it?

Ultimately, we shall concede that the answer is a qualified "yes." Anyone can operate an EMS the same way anyone can operate a racecar. There is an adage I use when explaining EMS operation and savings potential to customers and clients:

> *"The savings potential of an EMS is inversely proportional to the number of people who have access to it."*

If you give five people access to the EMS, you will have five people making changes. They may not tell each other what they did or why, they may not understand what they are doing. It will only take one of them to screw things up. The good news is that modern systems are very smart and are not likely to let you break anything or hurt anyone. They will positively ruin your savings very quickly, though.

As a customer and a building operator, you must understand your building and understand your systems and employ that knowledge to the management of your EMS. Otherwise, you will not see the savings you think you should.

This Really Happened...

I was working as a program administrator for a utility company located in the NorthEast, and my job was to evaluate the savings claims of vendors pursuing utility incentive money for energy-saving projects. If a project looked good and had good savings the utility would pay a portion of the project price to incentivize the customer to do the work.

One vendor submitted a project with a large ($410,000) EMS scope and was looking for approximately $150,000 in incentive (free) money. Furthermore, the customer in question was a large municipal facility dedicated to keeping individuals of a certain unsavory nature out of the public until such time as they had repaid their debt to society.

Now, my home (which is where I keep my family) was located about 20 miles up the highway from this customer, and as such I was highly motivated to keep this facility in tip-top condition. I looked forward to approving their incentive. Upon review however, I could not find the savings calculations or a detailed scope. The proposal had savings numbers but no supporting calcs. The scope essentially read, "Install EMS" with no other specifics. Confused, I called the vendor and inquired as to where I could get the calculations and details. The vendor, who was looking for a $150,000 handout so he could make a $410,000 project happen, said that it was a "standard calculation" and did not see fit to include it.

I assured him that nothing about this project was standard and that I would hold onto the $150,000 until the scope and savings could be quantified. I received the next day a very detailed scope of work, which was nice, and a spreadsheet with his calculations. Here is what the spreadsheet looked like:

Current Elec	Current Gas	Savings	Savings	Proposed	Proposed
kWh	Therms	Elec	Gas	kWh	Gas Therms
2,568,154	165,254	256,815	16,525	2,311,339	148,729
$ 359,542	$ 181,779	$ 35,954	$ 18,178	$ 323,587	$ 163,601
	SAVED	**$54,132**			

All he had done was multiply their current bills by 10% to achieve his savings number. I could not believe a professional in our industry could submit this for review with a straight face. And yet there it was, mocking me.

Of course, this was not the customer's fault, and I wanted to see the project happen. I calculated the savings myself, and surprisingly the savings were even higher than the vendor claimed.

This vendor put virtually no effort into determining the real savings potential of this project. None. He did not care at all whether or not the customer saw one thin dime of savings. That's pretty bad, but it's worse when you realize that this did not stop him from charging the customer $410,000. Want me to make this not-so-funny story less funny? This could be your vendor. He could be pitching you a project right now…

When are they a Good Idea?

Always. An EMS is a powerful tool. If you can afford one, get one. Get the best one you can afford. The key to saving energy is controlling the flow of energy to your and through your building. This is the best way to do it.

Just don't get caught up in advertising hype, and bad savings calculations. Know your building, know your systems, and go get those savings.

Myth #11: Power Conditioning Devices and the Science of Obfuscation

I apologize in advance for this chapter. I want this text to be accessible and useful to any building operator, regardless of trade, education, or technical know-how. But we have to talk about these magical black power-conditioning boxes that are getting sold to unsuspecting facility operators.

These devices get sold every day precisely because the underlying physics of what they do is really hard to understand. I made a joke earlier about Fourier's transformations[46] and this is where that joke comes from. This chapter will be painful for all of us. Many arguments and long whiteboard sessions went into the development of this chapter and my mother still openly weeps when it is mentioned. Let's all just resolve to do our best and get through this together. For all you signals and electrical engineers out there, I'm sorry if I butchered this too badly.

On with it:

There is a species of device, perching at the edge of the efficiency industry, that will often try to insinuate itself into the pantheon of ECMs that get included in other projects.

It goes by many names. It has many forms. It wears many disguises. It is legion.

It also rarely saves the average customer any energy. Which is a problem. We are talking about "power conditioning devices" or PCDs, as it were. They are often

---

[46] *The Fourier transform is called the frequency domain representation of the original signal. The term Fourier transform refers to both the frequency domain representation and the mathematical operation that associates the frequency domain representation to a function of time.*

marketed as harmonic balancing, impedance reactors, transient voltage suppression, and lots of other cool-sounding science-y names. How do they work? Well, just remember...you asked for it!

Problem #1: Power Factor. Not as Cool as it Sounds.

"Power Factor" sounds like something from a comic book or a heavy metal band name, which would be great. But it's far less interesting in real life. It's a little weirder in application but stay with me here.

Your power comes in from the pole at 60 cycles/sec (Hz) of alternating current. Basically, as the current "alternates," the power oscillates between positive and negative sixty times every second. That makes a nice, happy sine wave if you graph the power of a single phase as your meter sees it:

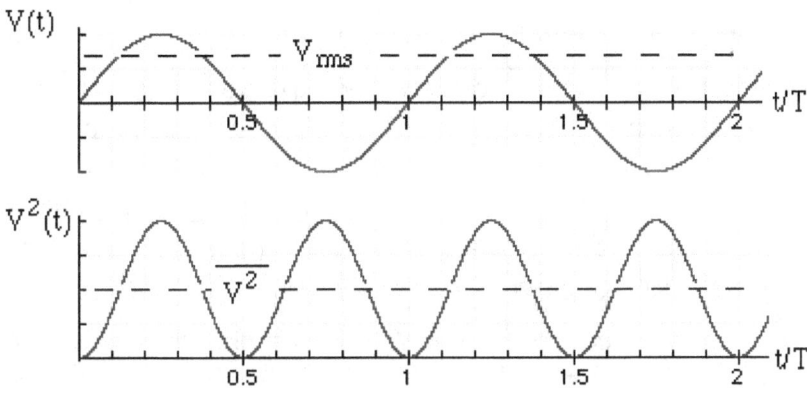

Top graph is the voltage passing zero.

"Three phase" power, is simply three of these waves spaced out at even intervals, and looks like this if you decide to waste time graphing it:

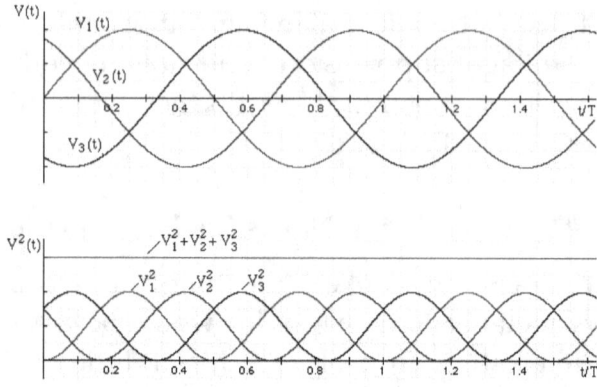

Every second, each phase of the power crosses zero 120 times. 0 is not a lot of voltage. 0 volts is bad if you want to do a lot of work.

Having three phases means that at any given time, two of your phases are NOT crossing zero, and therefore more usable power is available when compared to a single phase. Remember, at least 120 times a second alternating current at 60HZ will have a magnitude of "0".

When the waves are nicely synced up like in our graph, a lot of smooth usable power gets delivered to those power-hungry machines in your building. What happens when the waveform gets out of phase? More total power must be sent by the utility company to the devices so that the necessary kW is delivered.

The ratio of "real" power (kW they can charge you for) to "apparent"[47] power (the amount of kVA the utility has to make to get the total kW to you) is expressed as a dimensionless percentage and called "power factor."

---

[47] *These terms are in quotations because my father thinks "real" and "apparent" are terrible descriptors for this phenomenon. No one else does this, but he has decided they are wrong and that is that. This is pretty much a metaphor for my childhood.*

Power Factor (PF) by definition is the ratio of "real power" to "apparent power" in Volt-amps:

PF = Real power / Apparent Power
Or:
PF = Watts/VA

Why VA and not watts? Aren't watts just volts times amps? Basically, they needed a separate unit for the "apparent power" to differentiate the two is all. Something with a given watts rating may need more apparent power than real power, so a separate unit made life easier for everyone. Here is the important takeaway:

Any power factor quantity less than one means that the circuit must transport *more* current than what would be necessary to deliver the same quantity of real power.

Here is a little math. Picture, if you will, two motors.

Motor A is 10 BHP (7,460W) at 230v and has a power factor of 70%.
Thanks to Ohm's law, we know that:
Current (Amps) = W/ (Volts* Power factor)
By doing that elaborate engineer math[48] I can determine that the current required to give me my precious horsepower is:
Amps= (7,460W)/(230V*.7) = 46.3A
Motor B is the same, except that its power factor is .97...which is MUCH better. So, the math goes like this:
Amps = (7,460W)/(230V*.97) = 33.4A

That amounts to a 28% reduction in current (amps). Big savings!

---

[48] *"high school algebra"*

Hold up that parade, there, friend. It's about to get rainy. This situation manifests as small, but consistent losses between the power plant and the devices drawing the power, absolutely. So, to get your motor to run at nominal loading, the power plant has to send higher-than-nominal juice to your meter. **But the meter only reads the usable portion of the power, and not all the losses that the plant had to make up for**. You are only going to pay for the 7,460 watts either way. The power company has to make 12.9 extra amps (therefore 10,649 VA) to get you your 7,460 watts, but you only bought the watts. They don't charge for amps.

So, while the power company is forced to make a bunch of extra electricity that won't actually spin the meter, they typically can't charge you for that extra power. Which means for 90% of the accounts out there, this situation costs you, the consumer, NOTHING.

Your utility really hates it when it can't charge you for stuff. Charging you for stuff is sort of their thing. Thus, the modern facility with big inductive loads started to really wreck the power factor in industrial areas and was costing the utilities big dollars. Dollars they would like to recoup.

However, if you do not have consistently bad power factor, then the utility does not charge you anything at all for it. There is no savings to exploit here, so your PCD guy is barking up the wrong tree from the get-go. This will not discourage him at all.

If your power factor is consistently bad, then the utility may charge you extra for being a bad customer and a bad neighbor. (They don't actually call you a bad person, but it is implied.) Far be it from me to defend the utility charging practices, but this is a fairly legitimate gripe. In that case, you may find yourself paying for your bad power factor as a separate charge on your bill. Realistically most people won't have to deal with that. If you do, you will see it on your bill.

So here comes our PCD salesman with the solution to all your power factor problems. He has an expensive black box that will correct the power factor and save you energy!

If you are wondering exactly how this saves anyone energy except the utility company, then we are in the same boat. Your devices are all drawing exactly as much energy as they were before, and your meter is spinning at the exact same rate it always has. If you were getting charged extra for having a bad power factor, then you will feel some relief on your bill. That's a good thing. It is exactly why these devices were created in the first place.

Problem #2: I Squared R?

To convince you that their magic black box will save you energy, many vendors will wax eloquent about the losses associated with "I^2R" which is a formula for something boring and electrical[49]. It basically covers how much resistance to electrical flow your wire has. Which sounds weird because wire is supposed to conduct, not resist. If I am permitted to wax pedantic for a moment, I'll point out that technically speaking, everything has some electrical resistance. If that resistance is very, very, low, we just call it a "conductor."

Since even the low resistance of a conductor has to be accounted for when buildings are wired, a colloquial term for it was adopted. Your electrician may call it "copper loss," and it goes like this:

---

[49] *The pros call it "copper loss."*

Copper Loss: Explained

Copper losses result from Joule heating and so are also referred to as "I squared R losses", in reference to Joule's First Law. This states that the energy lost each second, or power, increases as the square of the current through the windings and in proportion to the electrical resistance of the conductors.

That's where we get I^2R from: "I" is the current flowing in the conductor and "R" is the resistance of the conductor. With I in amperes and R in ohms, the calculated power loss is given in watts.

Joule heating has a coefficient of performance of 1.0, meaning that every 1 watt of electrical power is converted to 1 Joule of heat. Therefore, the energy lost due to copper loss is:

Copper Loss= $I^2 *R*t$

Where t is the time in seconds the current is maintained.

Electricity is weird. When it flows through your wires, the higher current waveforms travel along the outside edge of the copper. Why? Because Faraday said so, that's why. The problem is that when you have a lot of extra non-useful current, it means less conductor for the current you actually can use.

Think of it as a highway with eight bustling lanes of intrepid electrons on their way to work. When everyone is following the rules and moving at the same speed, everybody gets to work on time. Now imagine you have a bunch of extra drivers who are driving too fast or too slow *and* not doing any work. These useless drivers are simultaneously taking up the outside lanes and not contributing the jobs you are trying to get done. This limits the number of lanes that the useful electrons can drive on. When the road has more of this extraneous traffic, fewer people get to work on time. When people can't get to work on time, they get angry and anger creates friction. Friction becomes heat, and heat changes the conductivity of the wire... In a bad way.

Poorly constructed analogies aside, the movement of electrons creates heat in proportion to the resistance of the wire. More resistance begets more heat and paradoxically, more heat begets more resistance. The hotter the wire gets, the less current it can move. The more current it moves, the hotter it gets. The easiest way to allow for this is to use bigger wire. More highway lanes means less congestion. Making it worse, these line losses and copper losses are multiplied by the length of the wire in question as well. It is an entirely intuitive to arrive at the conclusion that this might create an enormous opportunity for savings. Why would it not? Long wire runs will have larger losses, and modern buildings employ a whole lot of wiring.

Imagine how dizzy this situation makes your PCD sales rep!

Copper loss is one of several reasons selecting and sizing wire is its own course of study for the aspiring electrician and electrical engineer. The craftsman in question must size wire in compliance with electrical codes that often state that if the line losses are more than 5% the professional MUST go up to the next wire size. Within those constraints, he or she has to maximize the amount of current that can be moved without overheating the wire and creating resistance (and fire!), while also minimizing the use of expensive copper. If you have extra current on your wires because of bad power factor (or any other reason), you *will* have extra heat and extra copper loss. Your PCD rep will point this out and he or she will be right.

Now let's look at an example using our motor from the previous section, and try to figure out what sort of savings potential exists, shall we? We know that because of the bad power factor @ 70%, the utility has to make an extra 12.9 amps (compared to a power factor of 97%). You didn't pay for those amps, but that extra current is out there clogging up your wires and generating extra heat and resistance. If you don't want to do the math, just skip to the answer at the bottom of the box...

In our example, we have determined that the extra amps from bad power factor = 12.9A

I^2R = Watts lost

So:

$12.9A^2$ = 166.4W

Let's assume that the wire run is a staggering 200 feet down and back and the temperature of the conductor hits 75°C (167°F). Why such big and weird parameters? Because we want to give the PCD every chance in the world to be successful, here.

Our electrician is good at his job and specifies the correct wire size (#6), because lawsuits are awful.

#6 copper wire has the following resistance profile:[50]

#6 wire (65a, 75°C) = 0.3951Ω/1,000 ft. (or 0.0003951Ω/ft.)

At 200 ft. (down and back) the extra kW caused by those 12.9 amps amounts to the stunning quantity of 0.013 kW. Thirteen thousandths of one kW. 13 total watts. If, for some asinine reason, your motor ran 100% of the time, at $.12/kWh, the costs to you amount to the princely sum of:

8,760 hours = 115 kWh

$0.12 / kWh * 115 kWh = $13.80 / year

Now, your PCD vendor will tell you that all your other currents and transients are also heating up your wire and

---

[50] *Resistance is measured in "ohms" and the symbol is Ω.*

reducing its conductivity. He is not technically wrong, but the myth of magnitude was invented for this sort of thing. If your wires are sized properly, this will never amount to more than a few pennies a year. The men and women who do this for a living KNOW how much wire to use to make stuff work right. You have enough wire, I promise. If your wires aren't sized properly, then you have other, more pressing problems[51].

Objectively speaking, crowded wire cannot transport as much power as it used to because the highway of copper your electrons are driving down is jammed up with high-speed teenage electrons and blue-haired old-lady electrons going slowly in the passing lane with their turn signals on. To overcome this, more electrons have to be forced down the functional lanes. Am I beating this analogy too hard? The point is that transients, harmonics, and bent waveforms use up wire and don't do anything helpful. This makes you pull more current.

PCDs can and will fix this, and I can go into a lot of detail about it, or I can supersede the discussion and make one point:

The extent that this scenario affects your utility bill, however, is almost always completely negligible.

Your PCD will be installed *after* the meter. The utility owns your transformer and it is out at the pole or mounted on the ground. Either way, it's theirs not yours, so any corrections to your power factor must occur *after they have gone through your meter*. Which means that it can only correct energy consumption problems AFTER you have paid for them. Every surge, transient, blip and burp in the power has already passed through the meter before you

---

[51] *Like the fires and smoke. Also the lawsuits.*

can fix it. Any incidental increase in copper loss due to the extra heat is negligible at best and non-existent at worst.

But don't worry, just breathe deep and remember these things:

1. Your equipment is already pulling as much juice as it needs (you haven't changed that).
2. Your meter was only reading the real power anyway. None of this other stuff was affecting your bill.

All you can do with a PCD is clear up the highway for your electrons, which is nice. You already bought those electrons, though, and you still bought only as many as you needed to do the job.

Problem #3: But the Harmonics! And the Myth of Magnitude.

"Harmonics" refers to transient, unintentional waveforms that collect along the different phases of your system in such a manner that they amplify either each other or the "fundamental" (sinusoidal) current.[52]

Since we already know that random electrons moving around your wires at random frequencies bend stuff, what's different about harmonic distortions? If these rogue waveforms bend stuff in such a way that they "harmonize" with each other or the fundamental wave, then they are called "harmonic" distortions. Due to the nature of modern electrical distribution and its relationship with the number

---

[52] *Any current that has a frequency equal to a whole-number integer multiple of the fundamental frequency is harmonic.*

three[53], the most pernicious of these is the "3rd order harmonic" or "3rd harmonic.[54]"

How does this occur? It's because power often comes in on three phases, and not everything in the place uses all three phases. Many of your building's electrical devices are single-phase only, and that means the different phases coming in from your utility pole can have different loads and current distortions. In ye olden days of yore, there was not a whole lot of imbalance between the phases because there was not a whole of variety in equipment. A good electrical engineer could set it up so that all three phases were fairly well balanced and the problem would be pretty much handled. Furthermore, without modern sensitive electronics, harmonics just didn't cause that much trouble anyway, so even if he screwed it up badly...no one was likely to notice.

Nowadays, lots of phases can have lots of distortions. That's not too catastrophic by itself. It's when distorted currents in the phases are aligned such that all the positive peaks hit at the same time, and all the negative peaks hit at the same time, that they amplify each other.

Turn the book sideways for a second and take a look at this handy graphic.

---

[53] *Nikola Tesla was very OCD, and needed for everything around him to be divisible by three. No joke.*

[54] *Multiples of the 3rd harmonic (9th, 15th, 21st, etc.) are called "triplens" and are pretty bad themselves. The 5th harmonic, while not a triplen, is particularly troublesome for motors.*

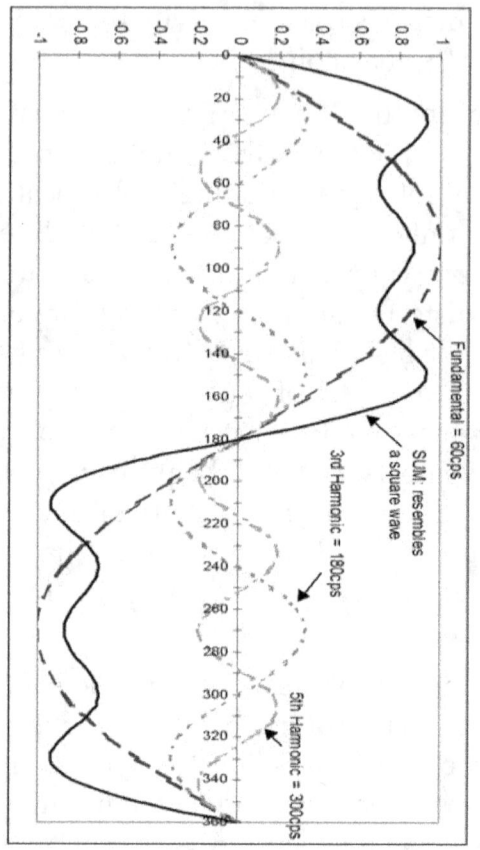

The tall dashed wave is the fundamental current. It's a pretty sine wave because that's how the utility sends the power. The smaller dotted lines are two different harmonic distortions. They are pretty too, I guess, but life is not a beauty pageant and we don't like them. The solid line is the sum of all of those currents on the fundamental current. Look at how it "flat-tops" the curve and reduces the peak current. We don't want that. We want clean sinusoidal power, not square waves.

When stray currents harmonize, the current *will* go somewhere (physics!), and ends up on the neutral wire of your electrical system. [55] This, in and of itself, is no big deal. It's part of the neutral wire's job and raison d'etre. It's where all the weird currents end up. All of the harmonics from all three phases go to that poor, beleaguered, neutral wire, and that can cause the current load on the neutral to get kind of big, and subsequently collect as heat. Special transformers are used in areas that have an issue with this, and a double neutral is standard procedure in most installations for exactly these reasons.[56] The heat gets dissipated away from the working phases, copper loss and I^2R losses are kept to a bare minimum, and life goes on.

This condition is usually caused by high-speed, switched, single-phase loads operating on multiple phases of a distributed 3-phase system. Don't know what those are? It's all those fancy computers and monitors in your office, folks. Since they are all doing different things requiring different power levels on a single phase of your system, the waveform on that phase can get all squiggly. There are inductive loads from lighting, capacitors in EVERYTHING, and any single-phase motor in the place that cycles on and off randomly is making it worse. Your refrigerator is not helping, either.

Before you start hyperventilating, remember everything we have already talked about:

---

[55] *Buildings with three-phase service will have a 4-wire system that includes a neutral wire. In a polyphase (usually three-phase) AC system, the neutral conductor is intended to have similar voltages to each of the other circuit conductors, but may carry very little current if the phases are balanced (no harmonics). Basically, it exists to collect random currents and keep the phases balanced.*

[56] *In times of yore, it was acceptable to size the neutral wire at half the phase capacity. Those days are long gone.*

1. Your equipment is only pulling as much juice as it needs (you haven't changed that).
2. Your meter was only reading the real power anyway. None of this other stuff was affecting your bill.
3. If your wiring was done competently, there is no real way these things can increase your bill.

To be completely fair, if you own your transformer or have transformers inside the building that step down the voltage, it is possible that harmonic distortions are actually costing some (tiny) dollars. It is also fair to point out that some of that collected current on the neutral is current you bought, but aren't using, and as such is technically wasted.

Products that isolate harmonics will not save any of this. Some products *prevent* 3rd-order harmonics from happening at all, and savings can be had under those conditions. So, let's consult the "Myth of Magnitude" and the "Myth of the Quantifiable" and see what we get.

It is true and safe to say that 3rd order harmonics build up faster and carry more current than other types of waveform distortions. They are the most dangerous and damaging and should be corrected in any facility with a lot of sensitive electronics. PCDs can and will fix this. They will either isolate the harmonic currents or prevent them from happening altogether. The only way to save *anything* is if you eliminate them.

So PCDs that eliminate the harmonics work, but working is not the problem. The problem is that as far as your utility bill is concerned, you are only eliminating the copper losses associated with the quantity of current that is correctable from your non-linear loads for the duration of the distortion itself.

That is a statement with a lot of conditionals. Five, to be exact. Each conditional reduces the savings potential of the device.

In the case of devices that prevent 3rd harmonics altogether, many vendors will demonstrate savings of up to 7% of your bill, and "prove" it with demonstrations and tests backing their claims. Typically, they will put a whole bunch of non-linear loads on a test circuit and graph the waveforms and amps before and after the PCD. Presto! 7% reduction in lost current!

What they neglect to mention is that the savings you see are for a load that is 100% non-linear, 100% of the time. Your building load is not entirely composed of non-linear loads. Many and most of your loads are either completely linear or are only non-linear for a few seconds at a time. Furthermore, just because a load is non-linear doesn't necessarily mean it will generate 3rd order harmonics. Without very thorough testing, you cannot know if a PCD will save you anything at all!

That's the Myth of the Quantifiable covered, so what about the Myth of Magnitude? If a PCD can eliminate all your 3rd harmonics, and that results in a 7% reduction in current associated with such, then great! But what percentage of your building loads are non-linear? If you don't know, don't feel bad: Nobody knows. You would have to buy a special meter and track it for a long time to get a good feel for your total harmonic distortion and its impacts on your equipment and energy bill. But for the sake of argument, let's assume 50% of your loads are non-linear and generating the harmonics associated with that. If you run a medium-sized office building, it is entirely possible that 50% (or less) of your loads are non-linear. Them computers is rough, man.

So you've got computers, but those computers don't run all night and weekend, do they?[57] No, they should be off or hibernating, greatly reducing your distortion for those hours. Even then, the total amount of distortion could be close to nil if the computers aren't doing much or your phases are well-balanced.

Either way, what you will actually save is 3-7% of the current from your non-linear load losses for only those hours that they actually distort the waveform. Consequently, it's more like a best-case scenario of 7% of 50% of your total load for maybe 25% of your annual hours. Or more succinctly, .875% of your current draw.

Maybe your computers never turn off. OK. 7% of 50% for 100% of the time is still only 3.5%. That is, if (and only if) your devices are truly creating the *maximum quantity of distortion possible for the maximum hours possible*. This is exceedingly rare. Like, "unicorn rare."

And by the way…YOU DON'T BUY CURRENT. YOU BUY POWER. The extra current can create extra heat, and as such cause a miniscule increase in the wire's resistance…but you now know that you are chasing fractions of a fractional fraction at that point.

The actual, average savings from these devices, under optimal conditions, is usually less than 1% of the total electric bill. It is confrontational to say it, but most PCD installations will save the average customer somewhere in the vicinity of 0%. Consider that carefully before you write that check.

---

[57] *If they do, turn them off! Save some real energy!*

This Really Happened...

Honestly, anyone in this industry for any length of time will have dozens of stories about the vendors of PCDs. This is one of the oldest and most pervasive efficiency snake-oil products around. My father oft speaks wistfully of the salesman who wanted him to consider the savings of such a device assuming the resistance of the copper wire was 1 ohm. Of course, the resistance of 200 feet of #6 copper wire is .0003951Ω/ft., or 39/100,000 of an ohm. You would need 2,564 ft. of that wire to make one whole ohm of resistance. It's just silly. If you need to inflate the baseline by 256,400% to make it work your product is bad.

Here is a fun fact, most utility companies will not even permit vendors or customers from applying for incentive money on any project utilizing a PCD. They love for you to correct power factor, but power factor correction is an established technology. For those customers getting charged for it, power factor correction does not require any fuzzy math to be a good idea.

That's right, virtually any version of power conditioning is automatically excluded from most energy efficiency incentive programs. The Massachusetts Technical Assessment Committee will not even pick up the phone for you. When I was submitting to the committee a (completely unrelated) product for peer review and utility incentive eligibility, it was the second question they asked me.

When is a PCD a Good Idea?

If you are paying for power factor, then go buy some power factor correction. That one is a no-brainer. Lots of reputable power factor correction companies and products exist. Go buy one.

Otherwise? These are a pretty tough sell. When you have a lot of sensitive equipment and/or a known issue with power quality, these can be a good investment. For many facilities, they are not merely a good idea, they are an essential component of the system. You will not likely find a data center or server farm without one or more. Important and sensitive laboratory equipment often benefits from meticulous power correction, as well. But realistically, they just don't save much energy or money unless your building is very, very, unique.

Certainly, all of the things that PCDs fix are things that can and will wreck modern sensitive electronics. To be fair, PCDs were not originally invented to trick people. They were invented because the old power distribution system was ruining modern technology. Your computerized equipment will probably last longer and run better with a PCD. You may even, under the most bizarre of circumstances, save a couple of bucks a year. But very few customers will ever save anything, let alone enough to justify the purchase of a PCD on efficiency alone.

Myth #12: O2 Trim

Boilers and furnaces need oxygen. To put it as plainly as possible, that's just how burning stuff works. Combustion (or 'burning') is a form of rapid oxidation, which not surprisingly, requires oxygen. All boilers (and furnaces) have to introduce fresh air to the fuel stream and combustion chamber to get the magic to happen. There is nothing mysterious or revolutionary about that.

The problem is that this fresh, oxygen-rich air will be colder than the flame it is fueling. In order for the oxidation to occur quickly (combustion!) that air has to get very hot. The chemical reactions that produce the heat/flame only occur when the air is super-hot. Which is why stuff just doesn't burn all the time.

Man Makes Fire!

Some materials have much lower thresholds for oxidation than others. Iron, for one, oxidizes at very low temperatures. Typically, this happens very slowly and we call it "rust." If you rapidly increase that rate of iron oxidation, then you will get sparks.

How does one do this? Early man did it by shaving some small pieces (with a lot of surface area) off some iron-rich rock with a piece of flint, causing the reaction to occur very quickly. Sound familiar?

When you strike a flint with a piece of steel, the iron atoms in the steel get stripped away in tiny pieces (with a lot of surface area); which end up in contact with the oxygen in the air. Subsequently, they oxidize very rapidly and ignite, which produces a lot of heat (spark). This kick-starts rapid oxidation (burning) that will continue as long as there is enough air and fuel to maintain the reaction. Now you can build that fire and survive another night in the frozen wilderness...

Since fuel does NOT oxidize at low temperatures, you have a problem. If the combustion air you have access to is 70°F, and the ignition temperature of fuel oil is 494°F, then your combustion air is going to have to rise from 70 to 494. That uses up BTUs that should be going into the water or airstream. It's really non-negotiable, and no device can fix that.

Furthermore, combustion air contains moisture that must also be heated, and then subsequently that heated water is carried up the stack and not into your building. Remember our condensing boilers? They steal that energy back by condensing the moisture out of the stack gas.

So, to ensure that enough oxygen is present to maintain the reaction, boilers are usually built to introduce about 15%[58] more air (3% more oxygen) than the combustion rate technically requires, simply to make sure that there is enough extra oxygen to ensure the complete combustion[59] of the fuel. This is referred to as "excess air."

Problem #1: Excess Air and the Myth of Magnitude

Generally speaking, many companies use a wide range of rule-of-thumb calculations to determine exactly how much savings potential O_2 trim has at a given location. These can range from 1% -2.5% "increase in efficiency" for every 1% decrease in excess O_2[60].

---

[58] *This is actually a code requirement in many states.*

[59] *Incomplete combustion on the other hand, results in carbon monoxide poisoning, soot, poor heat production, bad breath, poor credit, and general disapproval from your parents. Avoid it. It's the law.*

[60] *Amusingly, more than once a vendor has confused this to infer that a 1% decrease in COMBUSTION air correlated to a 1% increase in efficiency. Combustion air is only 20% O_2, causing the savings to be inflated by 500%.*

To accomplish this reduction, they will sell you a device that can detect the quantity of excess oxygen in your flue gas and trim the combustion air quantity to keep the O_2 level at optimum, eliminating a certain quantity of excess air. Excess air is cold and contains moisture and has to be heated by the burner to combust. More excess air will therefore adversely affect the combustion efficiency of your boiler. It's very intuitive.

It's also entirely too basic. The stoichiometry of combustion is a textbook unto itself, and any such "rule of thumb" predictions are going to deviate wildly over any significant sample size. And empirically, the number is more like .3-.5% increase in combustion efficiency for every 1% decrease in excess air.

So, in order to have real savings potential, the boiler plant in question needs to be using MORE than the state-minimum excess air (usually 15%), and enough more to justify the cost of installation. Let's do some math.

Assume a 1,000,000 BTU natural gas boiler with 20% excess air (4% excess O_2) is fully loaded for 1,000 hours. Because you are smart, you had a combustion efficiency test done, and know that your boiler is operating at 80% efficiency.

That means that you are buying 12,500 therms of natural gas for the year.

Annual Therms =
[(1,000,000 BtuH/80% eff) * 1,000 hrs]
100,000 Btu/therm
=12,500 therms

If you are paying \$1.10/therm, this is \$13,750 dollars a year.

Your vendor is going to install an O_2 trim system that will bring you to the state minimum of 15% excess air and 3%

Excess O_2. If we assume that it is realistic to expect .5% efficiency improvement per 1% reduction of excess air,[61] then the new math looks like this:

Efficiency improvement = .5% * (existing excess air %– proposed excess air %) or:

.5*(20-15) = .025 = 2.5% improvement.

If we add 2.5% to the existing 80% we get 82.5% combustion efficiency now. Plug the new efficiency into our equation above and you get a new gas consumption of 12,121 therms or $13,333. Your vendor has saved you $416 dollars. He'll likely want $10-15k for the widget, of course. But what's a 24-year payback among friends?

Problem #2: The Myth of the Quantifiable

How much excess air are you using? No one knows until they test for it. Of course, your vendor will be happy to do that test for you!

Dear reader, I have executed and reviewed many combustion efficiency tests. I am not a genius, nor am I the world's foremost expert on combustion. I **do** know that virtually every combustion test performed by a vendor selling O_2 trim has (shockingly) shown a large amount of excess air. Sometimes a ludicrous amount. Are they all lying? Of course not. There are plenty of bad boilers out there and there are a lot of ways to screw up a combustion test not attributable to dishonesty. But if you do not know with confidence where you are starting, then you cannot know what the savings potential is.

It's that simple. If you are considering this ECM, it would behoove you to have a competent, independent boiler technician come out and test the boiler. It takes a few

---

[61] *It's as good a guess as any and supported by decent research.*

minutes to a couple of hours, and you will get a result that looks like this:

```
- - - - - - - - - - - - - - - - - - - -
Log ID                          10
Date       14:07 03-Apr-2014
Fuel               Natural  gas

O2 (%)................3.5
CO (ppm).............49
CO2 (%)..............9.9
Ratio.............0.0005
Pressure (mBar).....5.59
Temp net (C).........40
Temp flue (C)........59
Eff gross (%).......88.3
XS air (%).........20.1
```

You can see that his boiler is fairly well tuned. The excess air (called "XS air" on the slip) looks high at 20.1%, but the O_2 is right at 3.5%, or damn near perfect. A vendor might give you this and say that he could improve your efficiency by 2.5% by bringing you down to 15% excess air, but there are two problems with that:

1. That would make the gross efficiency 90.8% (88.3+2.5%); which is a ludicrous expectation for a non-condensing boiler. This boiler is already running beautifully, even at 20.1% excess air. Why? *Because the stoichiometry of combustion is complicated!*
2. If 20.1% excess air equates to 3.54% excess O_2, and the relationship (on Earth) of O_2 to total air is always linear and constant, then reducing the excess air by 20% will reduce the O_2 by 20%, leaving it at an unsafe, illegal, and frankly stupid 2.8%.

Problem #3: There's a Better Way to do this...

Having just circumspectly accused the purveyors of O_2 trim devices of fraud and/or incompetence, I shall now backtrack by pointing out that in my own career I have more than once encountered boilers using 100%, even 200% excess air. I have seen and verified with my own eyes these boilers employing excess O_2 levels as high as 10-20-even 30%. At first blush these boilers certainly seem like very good candidates for an O_2 trim project.

Sort of.

Now, much like buying new windows is a bad way to fix leaky window frames, buying expensive and high-maintenance controls to trim excess air may not the best way to fix any excess air problems you may have. If your boiler is using too much combustion air, why not go ahead and give your local boiler maintenance company a call and have them tune your boiler for a couple hundred bucks?

Finally, if your boiler is well-balanced, well maintained, and is otherwise in good working order, you don't have that much excess air to trim. If you find you do have too much excess air, have a professional re-balance and re-commission your boiler instead. Maintenance is cheap, fancy controls of dubious use are expensive.

That's really all it takes for about 95% of the excess air problems floating around. You will achieve nearly all the savings of O_2 trim equipment with almost none of the (often considerable) expense. Furthermore, you have added no new and/or complicated devices to your equipment.

Problem #4: Don't Break Stuff or Kill Anyone, Please...

Due to the design of many types of modulating burner, plenty of burners are incapable of operating safely over variable fuel inputs without a lot of excess air. Tread carefully! If your burner has a high turndown rate and modulation, trimming excess air can be useless and potentially harmful.

There is a delay between when a molecule of fuel is burned and when the O_2 control gets to make a decision about O_2 levels. What will ultimately be the stack gasses do not quantum leap from the burner to the sensor, they will meander a bit and give their energy to the water/air or whatever. The sensor has to analyze this gas, and then send a control signal back to the damper or burner to make any necessary adjustments. A lot can happen between burning, sensing, and controlling.

Most systems employ a "trial and error" approach to maintaining excess air levels. They will perform a "trim" operation that opens or closes the combustion air damper, and then wait a prescribed delay period to determine the effect on the stack gasses. If the "trim" does not achieve the desired result, then the control performs another "trim." This process gets repeated until everything works within the desired parameters.

This works fine until the burner modulation control changes the firing rate even slightly at any point during this sequence.

And trust me, it will.

If the firing rate at the burner changes, the O_2 sensing and feedback loop have to restart. This will create a scenario where the control will be constantly "hunting" for the correct excess air setting, potentially starving and/or flooding the combustion process with excess air.

To prevent damage, explosions, deadly gasses and overall catastrophe, these controls are often commissioned to ensure that the boiler receives higher-than-necessary excess air while it waits for the next sample interval.

Wait a minute. Does that mean that the O_2 trim control is actually using too much excess air on purpose? Yes, it most certainly does. Equipment manufacturers don't like getting sued, it turns out.

More sophisticated systems interlock the O_2 sensors with the burner controls, which eliminates this problem, but those are the most expensive versions of this ECM. Most do not have this functionality.

When Does It Work?

There are few really good reasons to utilize this type of ECM.

1. Your boiler is huge.

These controls are fairly expensive for small boilers, but the savings with capacity scale faster than the costs. This ECM for a 20 million BTU boiler is not typically twenty times more expensive than a 1 million BTU boiler. Basically, the payback shrinks as the boiler gets bigger. There is absolutely a threshold whereupon these devices become solid ECMs with great savings.

2. You have emissions issues.

O_2 trim is a great way to control CO_2 and CO emissions, as they are all related to the completeness of combustion. Maintaining ideal combustion means cleaner stack gases, and a greener operation.

O_2 Trim is a classic case of a decent ECM being oversold. Like many others, under the right circumstances, it is a very solid part of your overall energy plan. Otherwise, it is a lot of sensitive expensive hardware that doesn't do any more good than regular maintenance.

Myth#13: Burner Pressure Boosters and Fuel Oil Atomizers Save 30%!

Years ago, when the end of WWII made oil cheap again and the American heating market began to abandon coal, there was a big shift to #2 oil for heating. At this time, almost all boilers used a fairly simple low-pressure injection system to feed the burner. Basically, oil was squirted into an igniter, set aflame, and blown through a heat exchanger or refractory. This method, when coupled with rudimentary cast-iron construction and single-pass design kept efficiencies legitimately low.

According to the "HEATING VENTILATING AIR CONDITIONING GUIDE 1952, " a standard residential or small commercial boiler (in 1952) could expect to see a total combustion efficiency of 60%-70%; which would be laughably poor compared to a modern device.

Then the 70's arrived, and a few grouchy oil exporters decided to let the world know that crude oil was now run by a cartel and not a free a marketplace. Oil prices soared, and suddenly efficiency became rather important. Several burner manufacturers developed a higher-pressure, smaller droplet producing spray nozzle. By pressurizing the fuel and spraying a finer mist, manufacturers were able to increase the total surface area of fuel available for oxidation. This amplified the rate of combustion and improved the overall quantity of fuel burnt per unit of total fuel mass.

When these improved nozzles were used in conjunction with superior heat exchangers and better overall designs efficiencies took a giant step forward.

Basically, if your droplets were big, then the combustion rate was slower, and more fuel ended up going up the stack unburnt. This unburnt fuel does not become usable heat and is therefore wasted. It was good science and a good idea…in 1970.

By 1980, further advancements and tweaks in boiler and burner design were achieved and it was expected that #2 fuel oil boiler would achieve 80% efficiency, and by the 90's 84-86% was the norm. If you have a condensing boiler, you could see efficiencies as high as 98% under optimal conditions.

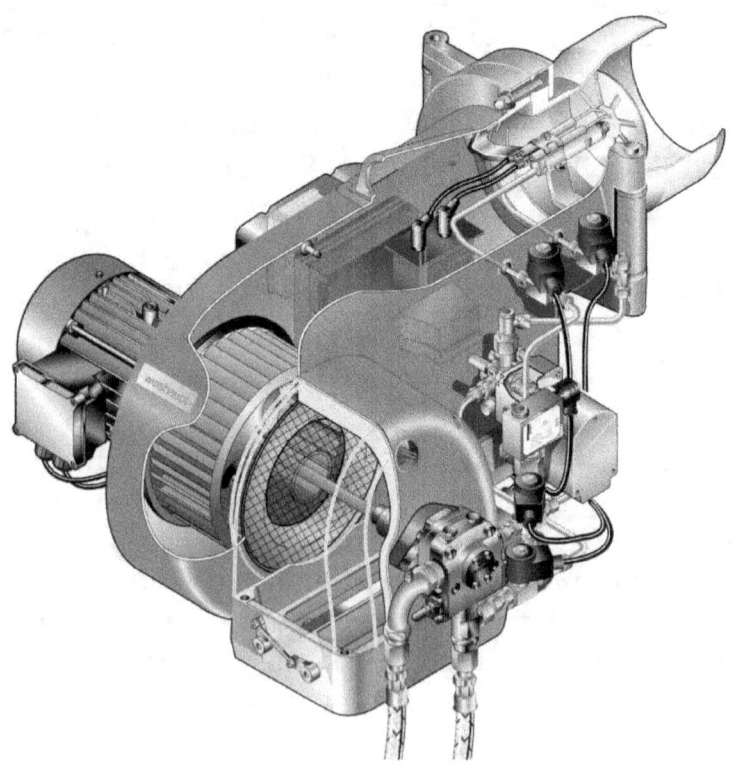

Look familiar?

The piece to take with your moving forward from here is this:

Virtually any #2 fuel oil boiler currently operating, provided it has not otherwise been neglected, will have no problem achieving 82-86% combustion efficiency.

Manufacturers of fuel oil boosters and atomizers capitalize on this by further reducing droplet size with very high pressures and very fine nozzles. The logic being that this would yield even higher combustion efficiencies.

Great. What does that mean?

"Efficiency" in this case means: "Heat energy out divided by fuel energy input." One gallon of #2 fuel oil has about 138,500 Btu of heat energy locked into its hydrocarbon bonds. If my boiler is 84% efficient and I burn a gallon of good ol' #2, I should get $138500 \times .84 = 116,340$ BTU out. Where did the rest of the Btus go?

1. They went up the stack as hot air.
2. They went up the stack as steam (because air has moisture in it, and you need air to burn stuff).
3. They became light (fire is bright).
4. They became sound. (fire makes noise)
5. They went up the stack as unburnt fuel. (It happens.)

It took 22,160 Btus to do accomplish these things, so I don't get to have them for space heating. That's life.

Problem #1: Misdirection and the Myth of the Intuitive

Many vendors of these types of products make some very extraordinary claims. Namely, 15-35% reduction in fuel consumption according to several prominent sellers. Extraordinary claims require extraordinary proof. How do they go about achieving this?

We understand the basic concepts at this point. Namely, by running the fuel oil through a high-pressure nozzle and reducing the droplet size we will increase the total surface area available for combustion. By definition, this increases the rate of combustion and reduces the quantity of unburnt fuel going up the stack. It also improves the fuel-to-air ratio to optimum levels during this process.

The following is some actual advertising material from one of these vendors. Paradoxically, they chose to not show any combustion efficiency testing, but instead point out that the flame achieves 1,700 degrees F using less fuel when their product is used.

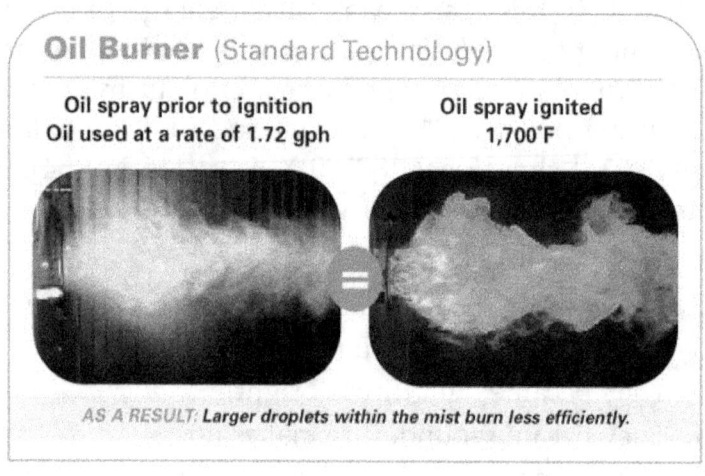

Oil Burner (Standard Technology)

Oil spray prior to ignition
Oil used at a rate of 1.72 gph

Oil spray ignited
1,700°F

AS A RESULT: *Larger droplets within the mist burn less efficiently.*

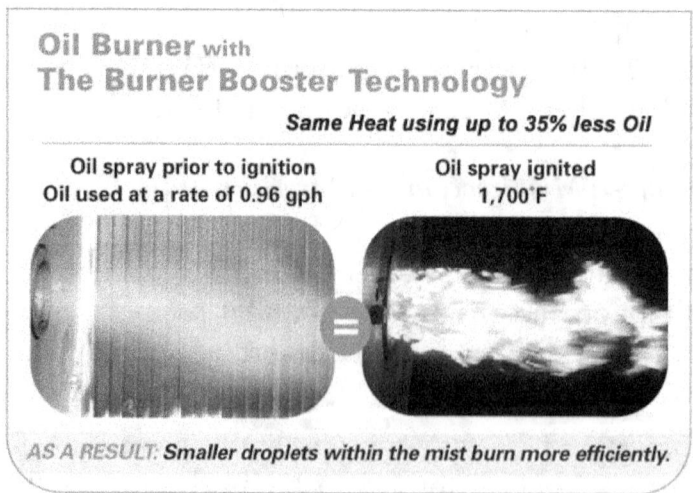

Oil Burner with
The Burner Booster Technology

Same Heat using up to 35% less Oil

Oil spray prior to ignition
Oil used at a rate of 0.96 gph

Oil spray ignited
1,700°F

AS A RESULT: *Smaller droplets within the mist burn more efficiently.*

Note that they claim the standard technology burns "less efficiently," and theirs burns "more efficiently." This will be important later.

Which sounds cool, until I point out that I can generate a 1,700-degree flame instantly with no oil at all:

My match is absolutely burning just as hot as that burner jet. Really. It is. Honestly, I'm not sure what they are getting at here; other than the fact that this is a classic misdirection.

What is important to your boiler is not the temperature of the flame, but how much of the fuel gets converted to HEAT. That is what "efficiency" is in this case. It is not "how much fuel it takes to get a flame of certain temperature," because that is an asinine metric. "How much fuel it takes to get a flame of certain temperature," has nothing to do with efficiency and everything to do with the chemistry of combustion. The head of the match is 1700 degrees as well; but I sure as hell can't heat my business with a match. The vendor is conflating temperature with heat output, and these are not the same thing. The match is the same temperature as the burner flame, but it sure as shootin' is not putting out as much heat energy.

Using this logic as ad copy requires us to accept one of two realities:

1. The vendor does not understand what he/she is talking about.
2. The vendor is deliberately spinning a tale to hide the weakness of their system.

Casting aspersions is a dangerous game.[62] So I will refrain from speaking to the vendor's actual intentions. This is a determination you will have to make as the customer.

Problem #2: Combustion Efficiency and the Myth of Magnitude

So, often a vendor will make claims like this:

"The [REDACTED] System reduces your use of heating oil by up to 35% on average."[63]

I'll just get right to it. That is a completely meaningless statement. How can something save "up to 35% on average?" That is like saying that my weight loss pill will take off up to 20lbs on average. It makes no sense. Let's have some fun:

- Up to 100% of people will die someday on average.
- Up to 50% of all teenagers will have acne on average.
- Up to 75% of politicians are lying, on average.

They are essentially claiming a massive swath of the statistical landscape as "average." The vendor's statement is almost always true, because if 90% of the vendor's

---

[62] *One of my favorite games, though.*

[63] *Real quote from real vendor's website.*

customers save .0005%, then it falls within the "up to 35%" designation.

The next sentence is even better:

"In fact, the Worcester Polytechnic Institute (WPI) Fire Science Lab has shown a reduction of just over 30% in the amount of fuel oil needed to produce the same heat, but this result is not typical or guaranteed."[64]

As your guide in this journey, I speak several languages, and in this case, I will translate from "Vender Hyperbole" into English:

"Under conditions we have not disclosed, with an experiment we won't tell you the particulars of, we managed to fabricate a scenario where we reduced the oil required to meet a load we won't elaborate on by 30%."[65]

Neither the particulars of the test nor a link to the research appear anywhere on their website. I dug up some of their ad copy that referenced the test, and it was far from definitive. Just the parts that they were willing to show us demonstrated that the methodology was poor and the test itself was admittedly riddled with equipment failures and poor calibration.

What almost all of these vendors are very careful to NOT say is that they are increasing *combustion efficiency.* They certainly imply that they can and will describe exactly that in any way they can that does not use the phrase "combustion efficiency" in the context of improvements.

---

[64] *Same Vendor. Same Product. Same silliness.*

[65] *I also speak "Salesman," "Lawyer," "Accountant," and "Angry Spouse."*

Why not? This is an amazing device. Right? Why not shout it from the rooftops?

They can't say that for one very simple reason: It's because combustion efficiency can be measured at any time by any competent boiler technician in the world. The WPI test they bring up was not a combustion efficiency test, which I found strange at first. Such a test would be the easiest and most categorical method of proving their claims. It would have shut my smug little mouth and killed this whole section of the book.

But what if it were true? What if they were reducing the fuel usage at the burner without improving the combustion efficiency? Unfortunately, as much fun as semantic gymnastics can be it simply is not possible to do this. Why is this?

Let's do a math problem.

- I have a building with 100,000 Btu/hr. of heating load. That means I need to put 100,000 Btu's of heat energy into the space every hour, or it will get cold in there.
- My boiler is well-maintained and properly tuned, so it remains a verified 85% efficient. This means that in order to get 100,000 Btu out of it every hour, I need to put 117,647 of Btu into it.

(100,000/85%= 117,647). That comes out to .85 gallons of fuel oil every hour (138,500 Btu in a gallon of oil).

Now, this "Burner Booster" vendor claims their device can reduce that consumption by 15%[66] (AT A MINIMUM!). That sounds awesome! So:

---

[66] *Actual claim. Actual vendor.*

.85 gallons – 15% = .72 gallons of oil per hour.

I haven't altered the efficiency, mind you, I've just reduced the quantity of oil required to meet the load.[67] What's so unbelievable about that?

Let's just calculate our combustion efficiency with the new consumption rate, then. Keep in mind, I haven't changed the LOAD. It is still 100,000 Btu/hr. But now instead of .85 gallon of oil per hour, I can achieve this with .72 gallons per hour. Same load, just 15% less fuel used to meet it. I haven't changed anything the vendor said, or taken them out of context at all, here.

If I achieve 100,000 Btu of heating with .72 gallons of oil, then my efficiency is now:

(.72 X 138500)/100000 = 1, or 100% combustion efficient!

This would mean that ALL of the available heat energy from the oil went into my heating water. Wait a minute, 100% of the thermal energy encapsulated in the oil molecules has now been transferred to the heating water?

That means:

- The flue gas temperature would be exactly the same as the intake temperature (room temperature), because none of the combustion energy heated the air between the blower and the stack.
- No water was boiled during the combustion process (somehow their device extracts atmospheric moisture and water entrained in the fuel oil?). Where is the reservoir or drain?

---

[67] *Yes, I know reducing the energy input required to meet a load is the exact same thing as increasing the efficiency. This is their silly claim, not mine!*

- There would be no unburnt or partially burnt hydrocarbons in the flue gas.
- The flame will have given off no sound.

Most of this is really easy to test. Try this: if a vendor stands by the claim of 15-30% improvement after installation, tell him he has to stick his hand in the flue when it's done. If he doesn't want to do that, ask him why. If he does do it, film that process and put it on YouTube. You will get 10 million hits overnight. You can call the clip, "Dedicated salesman burns his hand because he is a moron."

So, if I have an 85% efficient oil burner and I attach this device, the vender is essentially claiming that my efficiency will be 100% efficient (or more!!!)? Obviously, this is not the case. Achieving near-perfect combustion efficiency would be the sort of breakthrough that would have reverberated across the energy markets.

Simply increasing the pressure on the oil spray and reducing droplet size is neither new science nor revolutionary. It is just a question of diminishing returns. We are already squeezing just about all of the heat energy out of the oil molecules. A well-tuned burner will convert 99.98% of the fuel into heat already. Not a whole lot of room to improve there, is it? Where combustion heat gets wasted is in getting it to the medium (and eventually, the space) we want heated, not in the actual act of oxidation.

I think at this point we can be safely skeptical about the claims of this vendor. Obviously, we know that the device cannot possibly save us 15% if we have an otherwise functional boiler.

But what about that WPI test? Well, what they did was essentially a short-duration heat production bench test with very bizarre methodology and poor controls. The distinction is important, because if they claim they can produce the same heat for less fuel, then they are, BY DEFINITION, claiming to increase the COMBUSITON EFFICIENCY of the burner. We already know this claim is spurious. The BURNER is already converting 99.98% of the fuel to heat. Any boiler technician with a Bacharach Fyrite combustion tester will be able to tell you this. The BOILER however, is losing heat up the stack and through the housings. No pressure booster or atomizer is altering the stack or the housings, so they can't really change that.

For these reasons, vendors will go through significant verbal gymnastics to avoid saying that they increase combustion efficiency of the boiler; but alternative phrasing does not change the thermodynamics of the system. Here are some examples of how some vendors have phrased it, from various vendor websites:

- Creating a cleaner, more complete burn.
- What is normally produced as soot and harmful waste gas emissions is now turned into heat.
- Cuts the amount of heating oil used, by over 30%

In the "science"[68] section of one site, this little gem appears:

- "The [REDACTED] equipped heating system uses 15% - 25% less oil to produce the same amount of heat. "

All of these statements are versions of "Get more heat out of boiler with less fuel." *Which is the textbook definition of combustion efficiency. Why not just do a combustion efficiency test, then!?!?*

---

[68] *Their words...not mine!*

But, But...Testimonials!

Most of these vendors have several high-profile testimonials in their advertising. Testimonials are good ad copy; but testimonials are not data. Without extensive "before" testing, we have no idea what the (if anything) the product actually improved at any site. If a boiler has been neglected, and the burner badly out of tune, putting a brand-new burner on it can have a HUGE positive impact on the system. If excess air, fire rate, and nozzle geometry are bad, then the booster/atomizer will ABSOLUTLEY improve the efficiency...by a lot even!

Any of these sites could have had poorly maintained, very-old boilers. Replacing boiler is a massive expense. If an enthusiastic vendor was to show up and say they could save all that oil WITHOUT the outlay of new boiler, it might not take too much convincing to make that sale.

Of course, paying your service guy to maintain your boiler would accomplish the same thing. If you are in really bad shape, buying and installing a new standard burner will get you there too. Furthermore, vendors will often validate their claimed savings by adjusting usage using "heating degree-day"[69] data, which is popular with vendors because it is a generally terrible way to account for weather fluctuations.

So what is happening?

What is really going on when you get one of these? You end up with a brand-new burner, tuned and adjusted to perfection. If your old burner was in bad shape, then you are going to see some savings. That's a fact. Maybe even "a lot" of savings. Which is nice.

---

[69] *Heating degree day (HDD) is a measurement designed to reflect the demand for energy needed to heat a building. It is simplistic, problematic, and good only for broad estimating purposes.*

There is even some truth to the claim that the smaller droplets improve combustion. It's why we went to better spray nozzles in the first place. If I am to be fair, it would probably be safe and true to claim that these products can reduce *wasted* fuel by 30% in the context that they will reduce the quantity of unburnt fuel that goes of the stack.

Of course, after decades of R&D, burner manufacturers know how much pressure and what nozzle geometry it takes to get 999,800 parts of oil burnt per million. Said differently, they have already achieved 99.98% of available fuel burnt. Seriously, a modern, well-tuned oil burner will produce about 200 parts per million of unburnt fuel in the stack gas. To be fair to the vendors, I am willing to believe a product like this can cut that by 75%. So, if the standard burner is wasting .0002 gallons of oil per gallon burnt, then the Burner Booster will reduce that to .00005 gallons wasted per gallon burnt. Over a 3000-gallon heating season, your annual savings could be as high as 2 or 3 tenths of a gallon, annually. Or about 70 cents.

When are These a Good Idea?

If you really really really really really care a lot about optimizing your burner. Go ahead and pick one of these up. Otherwise, spend a fraction of that money keeping your existing burner in tip-top shape and put the rest back in your pocket.

Myth #14: Mystery (Snake) Oils

There are several versions of this one, but I am focusing here on refrigerant additives. These are expensive oils you can add to your aging refrigeration systems that will improve them and repair them and do other magical things that will save you lots of money and energy. Often referred to as "Polarized Refrigerant Oil Additives," (PROAs) these things are fairly commonplace in the HVAC marketplace.

They all make slightly different assertions, but the general thrust of the pitch goes like this:

Refrigeration oil is necessary to lubricate compressor parts and ensure a long and happy lifetime for the moving parts of your system. As equipment ages, and sealing barriers begin to underperform, some of this oil migrates throughout the whole system and can accumulate on the inner walls of coils and other heat exchanging surfaces. This is bad because oil is not as good a heat exchange medium as the copper (or whatever) that is supposed to be doing that job. Having a barrier of compressor lubricant between the coils and the refrigerant reduces the heat transfer between the space to be cooled and the coolant. All of that is a bad thing and none of it is necessarily untrue.

Manufacturers of these additives claim their product can eliminate these deposits by bonding to the walls of the system itself (preventing oil from sticking to it)[70]. The additive then sits there preventing lubricant form building up and keeps the whole system working nicely.

Claims of 10-30% improvement in efficiency are pretty easy to find.

---

[70] *My favorite bit of technobabble is from a vendor of one of these things touting the "nanomolecular bonding" of their product. It was awesome.*

Problem #1: The Oil Additive is Still Oil, and the Myth of the Intuitive

Right off the bat, the claim is not hard to understand. There is oil in the system, and oil sticks to stuff. When oil sticks to stuff, it doesn't transfer heat as well as it did before. If we alter that stuff so the oil doesn't stick to it, then we are doing a good thing.

The obvious question here is not whether or not the additive prevents oil buildup,[71] but whether or not having additive bonded to the walls is better than having oil bonded to them.

The National Institute of Standards and Technology (NIST) did some extensive testing and determined that the viscosity and density of most of these oil additives are no different from (or slightly higher than) standard refrigeration oil, which means that their heat transfer properties are also analogous to refrigerant oil.

Which begs the question: Why would having this particular oily substance stuck to the walls of your system be better than another oily substance?

From the Berkeley National Laboratory, we get the following:

> *"the viscosity of the additive is not greater than that of York oil-type "K", which according to the NIST theory eliminated the likelihood of enhancement."*[72]

---

[71] *Not a whole lot of proof it even does that…but I digress.*

[72] *"De-Scaling the Peak: The Impact of a Polarized Refrigerant Oil Additive on Chiller Performance at a U.S. Postal Service Processing & Distribution Center;" William Golove, Alex Lekov, and Gabrielle Wong-Parodi (Lawrence Berkeley National Laboratory)*

That is kind of on the nose there, isn't it? NIST and Berkeley Labs both agree categorically on this one.

But why would the manufacturer make a product that prevents oil buildup yet subsequently behaves exactly like the oil does?

That one is easy to answer. It is because your compressor does not like having chemicals in it that it was not designed to compress. Adding significant quantities of other media to a compressor that has been built and designed to compress refrigerant and a maybe little bit of oil will not end well for anyone. PROA vendors make darn sure that their product would not destroy the compressors it was supposed to help. These guys may be sneaky, but they are rarely stupid.

Having said that, HVAC techs for the most part hate the stuff. The chlorinated ones destroy compressors slowly, and the binders and other chemicals in these products clog filters and stick to everything. Basically, the good ones don't do anything, and the bad ones break compressors.

Problem #2: How Big is the Problem? (The Myth of the Quantifiable)

Once again, oil does in fact build up in older refrigeration systems. Especially so if they are not well-maintained. Here's the thing; all of these systems have filters and driers that exist for the express purpose of keep the refrigerant clean and contaminant-free.

If yours are in good working order, you will not have much buildup on your system. Period.

If you are operating a 19-year-old DX system that hasn't had a meaningful servicing in years? You probably have some buildup going on there, to be sure.

So, as an experienced and knowledgeable building operator, what is the best way to eliminate oil build-up in the system?[73]

1. An expensive oil additive of dubious merit.
2. A thorough assessment and cleaning from a skilled professional.

It's hard to say which will be cheaper. It depends on how bad your system is. But only one of these will actually locate, assess, and correct system problems. Your tech can easily and comparatively cheaply flush the system, clean it, and recharge it without having to rely on dubious claims of product efficacy.

Problem 3: Data Exists. It Ain't Complimentary

I am going to quote from a few independent tests, here. Just for fun…

Here is the Solar Energy Center talking about the manufacturer's supporting test data in a scathing meta-analysis of collected test data from all over the USA: [74]

> *"These reports describe energy monitoring and claim energy savings ranging from 12 to 36% after the COA was added. These tests measured "before" and "after" energy use, however none of these tests measured the load on the air conditioning or refrigeration equipment, that is, none of the tests made sufficient measurement whereby the*

---

[73] *Oil build up that may or may not even exist.*

[74] *"A STUDY TO DETERMINE THE ENERGY IMPACT OF ADDING POLARSHIELD TO AIR CONDITIONING SYSTEMS;" Charles J. Cromer, Ph. D, P.E., Program Director, Florida Solar Energy Center, Cocoa, Florida*

conditions under which the systems were operating for the before and after periods could be adequately determined"

Here they are in the same report discussing the findings of an Oak Ridge National Laboratories test (emphasis mine):

*"Table 2 indicates that **no improvement in heat pump performance was measured** in our laboratory tests as a result of adding this product to our test unit. The small changes (±2%) in steady state compressor power consumption and cooling capacity shown in Table 2 are most likely attributable to random experimental errors, although a small 2.5% improvement in EER is indicated for 3 ounces of the additive. It is also worth noting that the evaporator air entering/leaving temperatures and the compressor pressure ratio **showed no significant change as a result of additive addition**. Both of these observations are consistent with **no improvement in heat exchanger performance**. There was, however, a noticeable, but unquantified, decrease of compressor noise resulting from additive addition."*

Here is the discussion of their own 12-day steady-state testing.

*"The Unit #1 test results **show no improvement in the operation of the equipment and no energy savings that can be attributed to the addition of the COA**. But, Unit #1 did not operate worse after the COA addition. This system just operated the same within the accuracy of the test. However, there is a trend for Unit #2 that shows the unit operated a little worse, less cooling and lower EER when the first addition, 2.5 oz of COA was added. This trend is proved at 95% confidence with the second 2.5 oz addition, totaling 5.0 oz of COA added. The*

system used more energy, provided less cooling, and showed a lower EER. This is consistent with work from ASHRAE that indicates that too much oil in a system can degrade its performance."

Here is a tidbit from another highly-controlled test[75] of a PROA:

*"As summarized in Table 6.7, **there was a statistically insignificant change in operational efficiency of the liquid chiller** at or below 0.5% between test periods depending on the loading of the unit."*

*"For the tested conditions of the liquid chiller in question, it can be stated with sufficient confidence that the effect of the polarized refrigerant oil additive, known as [REDACTED), **was statistically insignificant with respect to liquid chiller efficiency** augmentation."*

I could literally fill thirty more pages with this stuff. I don't have to spend too much time debunking this one because EVERYBODY ELSE ALREADY HAS!

This is a product that has been *very* thoroughly analyzed. And it is at best, snake oil.

When is a PROA a Good Idea?

Never. The only thing that should ever be added to a refrigeration system is refrigerant and the manufacturer's approved lubricant. Period.

---

[75] CHESTNUT, CHRISTOPHER BLAKE; *"Experimental and Investigative Analysis on the Effect of a Polarized Refrigerant Oil Additive on Water Cooled Liquid Chiller Performance."* (Under the direction of Dr. Stephen D. Terry.)

If your system is old and performing poorly, then you need to fix it or replace it, not roll the dice on a magical elixir.

Myth #15: You Can't Be Serious...

Let's get weird for a second here. There is a lot of strange stuff out there. There is no shortage of "black box" devices that magically do...something... to make you use less energy. There are widgets that alter magnetic fields, rare-earth elements that increase efficiency, and my personal favorite, anything with "Tesla[76]" in the name. We are going to quickly go through some of them to give you an idea of what is and is not a good idea.

Magic Magnets

People love magnets. But that fascination has allowed them to be incorporated into many fields of pseudoscience. Here is a couple of good ones to watch out for.

Water Treatment

Electromagnetic water treatment has been touted as an inexpensive way to treat refrigeration and HVAC water loops to prevent scale buildup and otherwise improve water system efficiency. A device is attached to the exterior of a water pipe and strong magnetic fields are pulsed through the water as it flows by the device. There is precisely zero reliable evidence that these devices do anything at all.

Fossil Fuel Magnets

Multiple vendors currently offer various versions of magnetized or "rare earth metal" devices that purport to improve fossil fuel combustion. They claim that magnetic fields align hydrocarbon molecules in such a manner that

---

[76] *Except the cars. Love those things.*

oxidation occurs in a more uniform manner. Of course, hydrocarbon molecules are non-polar. Magnets and magnetic fields do not influence their behavior in any measurable way.

Filthy Fuel Fixers

Fuel Catalyzers

Another popular item is a fuel catalyzer. This device will be attached in your fuel line (heating oil, gas, whatever) and as the fuel passes through it, some sort of process occurs whereby the fuel molecules either align, or get broken down, or both. It will be either a magnet, or a special filter, a proprietary process, or some other vaguely science-y mechanism that accomplishes this. This then allows for more complete combustion and better more efficient heating. Sound familiar?

There are hundreds of versions of this and they all have one thing in common: they don't do anything. Hydrocarbons arrive at your facility in a highly refined state and they already combust pretty darn well.

These vendors will often have very convincing testimonials and even some "independent testing" to back up their claims. I can save you a lot of legwork on this. The tests are poorly executed, and the testimonials are cherry-picked. Walk away from these.

Here's the Deal

Innovation and experimentation is great. We all want to see new and exciting solutions to the problems of energy efficiency. Unfortunately, this breeds a certain recklessness among those who truly want to help and a certain ruthlessness in those who just want to make a profit. To defend yourself, remember these rules:

Honest people show their work.

If your vendor is secretive or hides behind "proprietary" data or methods, then go ahead and walk away. There is a pervasive mythology in this industry that there are certain "silver bullets" or breakthrough techniques that are jealously guarded trade secrets. This is just not so.

As a species, we have a fairly good handle on energy and how we use it. As we improve our understanding we make incremental improvements in the devices and techniques we employ. If a truly revolutionary technique is discovered that dramatically increases the efficiency of something there will be markedly robust examination and discussion of such. It will be researchable, reproducible, and documented. It will not be a trade secret.

Reputable vendors will be anxious for you to see this. They will provide third-party tests, on-site demonstrations, and robust calculations and peer-reviewed research to support it. They will want you to know how great it is. It might be worth your time to hire an expert to help interpret these claims, but either way good engineering does not shy away from scrutiny. If someone really has made a Silver Bullet, they will not be shy about letting you test it.

If it sounds too good to be true, it probably is.

Extraordinary claims require extraordinary proof. If the simple payback is less than a year, or the savings are more than 20%, it's time to be nervous. That's really the bottom line. It doesn't automatically mean that an ECM is bad or a vendor is lying. But large savings are very hard to achieve with small dollars.

How will you know? Take a good look at the before and after conditions. Did the vendor assume too many run hours or too much load?

Are the savings greater than quantity of energy consumed? You may scoff, but this is extremely common. A vendor employing less-than-robust calculation methods will often inadvertently predict savings in excess of the original load.

Testimonials are not data.

Heavy reliance on testimonials is like bright colors on an Amazonian tree frog: its nature's way of telling you to stay away. This is probably the number one method of spotting unreliable vendor claims. Logically, if the science of your claim is irrefutable, then testimonials are completely irrelevant. Facts do not require endorsement, after all.

Conversely, testimonials only require that one person to be happy with a product, which does not correlate to the veracity of the claims. I have encountered more than one client who felt that an ECM saved them on their heating bills, only to point out that the previous winter had been much colder than average. The heating bills were less because the subsequent winter was warmer, not because the ECM worked as advertised.

Sadly, vendors who rely on a lot of testimonials often do so because the facts often don't support their claims.

Savings Guarantees are neither savings nor guarantees.

Some vendors will guarantee savings from their ECMs, which sounds perfect. If it doesn't perform, you get your money back, right? Having reviewed many of these guarantees, I must sadly report that getting the vendor to pay on a guarantee is nearly impossible. Simply because virtually every savings guarantee out there places the burden of proof on the customer.

Most have wildly untenable requirements for said proof. Typically, the customer will have to monitor and measure for two years after installation. The customer must demonstrate that the building loads have not changed at all from the baseline as well. So, if you changed the coffee maker in the break room, you have changed the baseline, and voided your guarantee. Did you hire two more guys to work the warehouse? Whoops.

If you manage to get past all of that, then your lawyers will argue with their lawyers over what constituted savings and what the baseline really was. Then you can discuss whether or not your method of measuring was accurate, or if the weather affected the savings and by how much.

This is not always the case, obviously. There are reputable companies offering guarantees and doing so in a forthright manner. Reputable companies, with honest guarantees are not terribly difficult to spot:

- Their savings numbers will not be huge.
- They will have a plan for monitoring and verification (M&V) of savings.
- They will place the burden of proof upon themselves.

One advantage of a reputable savings guarantee is that the vendor will likely have purchased an insurance policy on it. Ask about that, because that is a good thing to know about. The underwriter is not going to be happy writing policies for bad products or measures. This translates into at least one objective and robust review of the project from someone else who has something to lose.

Ultimately, never forget that offering a savings guarantee is a sales tactic even for a reputable company. It is not necessarily a measure of confidence. Do not rely on guarantees from the vendor unless you are very good at spotting all the potential pitfalls.

Trust no one.

I know. It sounds terrible, but you should never trust a vendor. Not because you are jaded or because everyone is lying, but because trust is subjective. You are a practical engineer now, so you don't get to be subjective any more.

That vendor does not know your building as well as you do. The salesman doesn't know your product, or your employees, or your needs. He or she can't. Only you can know these things. I'll go one further: it's not even the salesman's job to know what you need. It's yours. The salesman's job is to sell you something. A good salesman will do their best to make sure you get what you need, and a bad one won't care.

So don't trust that the vendor will know what to do. Learn what you need for yourself, and then go buy it.

If you walked into a car dealership, and told the first salesperson who came up to you that you wanted a new car, but all you could articulate was a need for four tires and an engine, how do you suppose that salesperson might react to that? You might as well be wandering through wolf country with a pocket full of bacon. You have "nail" written all over you at that point and the hammers will be everywhere.

Now go to that dealership and say to the salesperson, "I would like a black 2016 Cadillac CTS-4, with the Luxury trim package, premium sound, and the ES3 suspension option. I have my loan already worked out and here is a list of all the prices listed for this car from all the competing dealerships in the area."

That, I promise, is two very different car-buying experiences. Why would you treat an energy project any differently? If you need to, hire a reputable consultant (with no connections to the vendor) to help.

If this sounds unreasonably cold and rigid to you, then you are reading it right. This is not just your capital money; this is your energy bill going forward. You have the right to demand that the vendor justify not only the purchase price, but the potential lost costs of energy bills going forward. If they can't or won't, go find one who will. There are many upright vendors and companies out there that will secure your business in a forthright and honest manner.

End of the Line...

There are many reasons to pursue energy efficiency. All of them are good reasons. Maybe you are looking to reduce costs. Good plan. Energy is expensive (and looks to remain that way for the foreseeable future). Or maybe your company is environmentally conscious and wants to engage proactively in reducing consumption and emissions. Bravo! The world needs companies to take a stand against environmental damage.

No matter what the impetus for your desire to increase your efficiency, the current climate in the efficiency industry has led to the proliferation of some less-than-accurate claims.

So, there it is. An industry laid bare, and its secrets exposed. For what? Hopefully for progress. This is an industry in flux, as the techniques and technologies evolve faster than they ever have before. It is an exciting time, but also dangerous and confusing. Even the professionals plying their trade in this environment get caught by some of the problems this book discusses, and no one wants to admit they were wrong.[77] We are all doing the best we can with what we've got. Being more energy efficient is a worthy goal and deserving of some time and effort.

With any luck, this text will help building operators avoid expending time and capital on things that don't help. Hopefully, it will help spur vigorous dialogue between vendors and consumers in the pursuit of the best possible solutions to the problems encountered in today's commercial buildings.

---

[77] *I'M NEVER WRONG. But if I was, I wouldn't admit it.*

Essentially, I want you to be loaded up with the best old-fashioned lead bullets your money can buy.

If nothing else, I hope that it educated and entertained.

INDEX